T0139038

The Theory of Everything

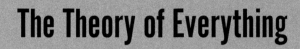

The Theory of Everything

Quantum and Relativity is Everywhere – A Fermat Universe

Norbert Schwarzer

JENNY STANFORD
PUBLISHING

Published by

Jenny Stanford Publishing Pte. Ltd.
Level 34, Centennial Tower
3 Temasek Avenue
Singapore 039190

Email: editorial@jennystanford.com
Web: www.jennystanford.com

British Library Cataloguing-in-Publication Data
A catalogue record for this book is available from the British Library.

The Theory of Everything: Quantum and Relativity is Everywhere – A Fermat Universe

Copyright © 2020 Jenny Stanford Publishing Pte. Ltd.

ISBN 978-981-4774-47-5 (Hardcover)
ISBN 978-1-315-09975-0 (eBook)

Meinen Eltern, meiner Frau und unseren 4 Kindern

Contents

About Motivation and Luck

While working as a consultant for the industry and especially when analyzing and evaluating certain measures in economics and politics, the author was always forced to observe the total uncertainty budget of his own models in a very critical manner. Almost nobody doubts a result coming out of a simple equation like 2+2, but things are quite different when models become more complex. This is even more so when the object of investigation has something to do with humans or when it concerns the human society as a whole. A simple example is market studies and market predictions. Coming up with a certain result when doing this, one better make sure that the error bars, which is to say the total model's uncertainty, are not as big as, or even bigger than, the result itself. This in principle holds for all models, of course, but even more so for problems where the researcher herself or himself could be or become part of the equation. There are numerous examples from politics, economy and sociology where bad decisions have been made solely because the decision maker was not able to distinguish between knowledge and ideology, between independent parameters or degrees of freedom and his own personal interests. For all these cases it would be advisable to have an—almost—unchallengeable method to account for the total, holistic uncertainty budget of the model or simulation, including the potential influence of the model creator himself or herself.

What could be closer to deal with this principle problem concerning all fields of science, development, research, prediction-making, investigation and so on, than to apply the very method or theory which only exists because of those principle uncertainties residing in our universe. We are talking, of course, about quantum theory.

In order to holistically and most generally solve that problem, however, we need to be able to quantize, which is to say quantum-theoretically treat, every mathematical model we intend to put under such a strict regime. Considering such mathematical models as mathematical spaces, the whole task just comes down to the problem of generally quantizing such spaces. As for most applications the mathematical models of a given problem can be formulated in a smooth, which is to say differentiable, way, it should totally suffice to have a quantization method ready for such smooth spaces. It was clear to the author that any attempt for such a general achievement automatically also solves the problem of quantizing the Einstein field equations. This, however, has been referred to as the Theory of Everything by the popular press. The author is quite critical towards this expression because, in his opinion, it is rather daring. Thus, originally, he intended to give the book a less "front-page-like" title. However, he was convinced otherwise by the publisher, because, even though quite inflationary, the title *The Theory of Everything* in fact covers the principle aspects of the approach, and this is even more so if we consider the origin of the socioeconomic motivation which was the reason to start this work at all.

Fully aware of what was lying in front, the author would probably never have started research if it were not for a crucial hint he had received from a book [1] at the right moment. There, in *The Eighth Day*, an equation has been given, which, as the story goes, was left by two Jews before they were killed by Nazis. Correctly interpreted, this equation opens up a wholly new way to expand the line element of any metric space, and applying this expansion, subsequent quantization appeared as a mere technicality rather than the "problem of the century," as the present author has called it in some of his lectures. Although it might seem rather controversial to some science purists that the author has taken ideas from potentially fictional-historic literature, the author himself merely sees it as luck to have got the right hint at the right time, and he does not care much about where it came from.

Chapter 1

Brief Introduction

The Stamler Approach: A Brief Historical Overview of the Original Idea

Stamler et al. [2] have shown how the assumption of a substructured space automatically leads to a quantization of the Einstein–Hilbert action [3, 4]. The classical Einstein–Hilbert action is given as

$$\delta_g W = 0 = \delta_g \int_V d^n x \left(\sqrt{-g}\, [R - 2\kappa\, L_\mathrm{M}] \right), \tag{1}$$

where g denotes the determinant of the metric tensor, W and V give the action and the volume of the n-dimensional space, respectively, R denotes the Ricci scalar, and the last term, $2\kappa\, L_\mathrm{M}$, describes the Lagrange density of matter. This action avoids higher orders of curvature—that is, the higher orders for the Ricci scalar and/or the metric (within the Lagrangian approach as used by Hilbert [4]). Evaluation gives the well-known Einstein field equations in n dimensions with the indices α and β running from 1 to n:

$$R^{\alpha\beta} - \frac{1}{2} R g^{\alpha\beta} + \Lambda g^{\alpha\beta} = -\kappa\, T^{\alpha\beta}. \tag{2}$$

Here, we have $R^{\alpha\beta}$, $T^{\alpha\beta}$ the Ricci and the energy momentum tensor, respectively, while the parameters Λ and κ are constants (usually called the cosmological and the coupling constant, respectively).

The Theory of Everything: Quantum and Relativity is Everywhere – A Fermat Universe
Norbert Schwarzer
Copyright © 2020 Jenny Stanford Publishing Pte. Ltd.
ISBN 978-981-4774-47-5 (Hardcover), 978-1-315-09975-0 (eBook)
www.jennystanford.com

Now Stamler et al. [2], as mentioned above, have simply applied a new form of the Dirac concept [5] on the metric gg_{ij} and constructed it as

$$gg_{\alpha\beta} = S_{\alpha\beta\mu}S_{\alpha\beta}^{\mu} \tag{3}$$

with S been defined as

$$S_{\alpha\beta\mu} =$$

$$\begin{pmatrix} \sqrt{g_{00}} \cdot s_\mu \, [t, tt, t] & \sqrt{g_{10}} \cdot s_\mu \, [t, tx, x] & \sqrt{g_{20}} \cdot s_\mu \, [t, yt, y] & \sqrt{g_{30}} \cdot s_\mu \, [t, zt, z] \\ \sqrt{g_{01}} \cdot s_\mu \, [x, xt, t] & \sqrt{g_{11}} \cdot s_\mu \, [x, xx, x] & \sqrt{g_{21}} \cdot s_\mu \, [x, xy, y] & \sqrt{g_{31}} \cdot s_\mu \, [x, xz, z] \\ \sqrt{g_{02}} \cdot s_\mu \, [y, yt, t] & \sqrt{g_{12}} \cdot s_\mu \, [y, yz, z] & \sqrt{g_{22}} \cdot s_\mu \, [y, yy, y] & \sqrt{g_{32}} \cdot s_\mu \, [y, yz, z] \\ \sqrt{g_{03}} \cdot s_\mu \, [z, zt, t] & \sqrt{g_{13}} \cdot s_\mu \, [z, zx, x] & \sqrt{g_{23}} \cdot s_\mu \, [z, zy, y] & \sqrt{g_{33}} \cdot s_\mu \, [z, zz, z] \end{pmatrix} \tag{4}$$

(with the components under the square root being those of the classical solution) and the 4-vector s_μ being rescaled to (for all arguments of $s_\mu = s_\mu \, [\alpha, \gamma, \beta]$)

$$s_\mu \, [x, xx, t] =$$

$$\frac{\left(\sqrt{t^2 + (x - xx)^2}, \ \sqrt{t^2 + (x + xx)^2}, \ \sqrt{t^2 + (x - i \cdot xx)^2}, \ \sqrt{t^2 + (x + i \cdot xx)^2} \right)}{\sqrt{4 \cdot (x^2 + t^2)}}. \tag{5}$$

Now, the present author has used this concept to quantize the line element in a simple "Dirac-like" manner [6]:

$$ds^2 = g_{ij}dx^i dx^j = \frac{\partial x^m}{\partial y^i}\alpha_m \frac{\partial x^n}{\partial y^j}\alpha_n dy^i dy^j = \frac{\partial \, [\tilde{x}^m \tilde{\alpha}_m]}{\partial y^i}dy^i \frac{\partial \, [\tilde{x}^n \tilde{\alpha}_n]}{\partial y^j}dy^j ;$$

$$\Rightarrow \quad ds^2 = g^{ij}dx_i dx_j = g^{mi}\frac{\partial f}{\partial y^i}dy_m g^{nj}\frac{\partial f}{\partial y^j}dy_n = g^{mi}\frac{\partial f}{\partial y^i}e_m g^{nj}\frac{\partial f}{\partial y^j}e_n dy^2$$

$$= g^{mi}\frac{\partial f_m}{\partial y^i}g^{nj}\frac{\partial f_n}{\partial y^j}dy^2$$

with e_m unit base vector here used as $dy_m = e_m dy$ $\qquad$ (6)

allowing to incorporate the Einstein field equation without any approximation. The resulting "quantum field" equation can be given as follows:

$$0 = \Upsilon \left(\left[R^{mi} + \kappa T^{mi} \right]\frac{\partial f_m}{\partial y^i}, \ i \cdot \left[\frac{1}{2}R - \Lambda \right]\sqrt{\lambda^m} \cdot f_m, \ \pm I \cdot \left[\frac{1}{2}R - \Lambda \right]\varepsilon^m f_m \right)$$

with λ^m eigenvalues of $g^{\alpha\beta} = \dfrac{R^{\alpha\beta} + \kappa T^{\alpha\beta}}{\frac{1}{2}R - \Lambda}$. $\qquad$ (7)

Here the symbol $\Upsilon\left(a^i b_i, \pm I \cdot c^j d_j\right)$ stands for an n-tuple expansion, which can be given as

$$
\left\{\left[\begin{array}{l}
i\left(g^{00}\partial_0 f_0 + g^{10}\partial_0 f_1 + g^{20}\partial_0 f_2 + g^{30}\partial_0 f_3\right) \\
+i\left(g^{01}\partial_1 f_0 + g^{11}\partial_1 f_1 + g^{21}\partial_1 f_2 + g^{31}\partial_1 f_3\right) \\
+i\left(g^{02}\partial_2 f_0 + g^{12}\partial_2 f_1 + g^{22}\partial_2 f_2 + g^{32}\partial_2 f_3\right) \\
+i\left(g^{03}\partial_3 f_0 + g^{13}\partial_3 f_1 + g^{23}\partial_3 f_2 + g^{33}\partial_3 f_3\right) - \sqrt{\lambda^m}\cdot f_m
\end{array}\right]_{\Omega=\Omega_1} \equiv [\ldots]_{\forall\varphi_{(mj)}=0},\ldots \right.
$$

$$
= [\ldots]_{\forall\varphi_{(mj)}=0},\ [\ldots],\ [\ldots]_{\forall A_{\varphi_{(mj)}}/\varphi_{(00)}=0},\ [\ldots]_{\forall A_{\varphi_{(mj)}}/\{\varphi_{(00)},\varphi_{(10)}\}=0},\ldots,\ [\ldots]_{\forall A_{\varphi_{(mj)}}/\{\varphi_{(00)}\ldots\varphi_{(kl)}\}=0}
$$

$$
\varphi_{(00)} = \pi \qquad \varphi_{(00)} = \varphi_{(10)} = \pi \qquad \{\varphi_{(00)}\ldots\varphi_{(kl)}\} = \pi
$$

$$
\varphi_{\Omega(mj)} = \{0,\pi\}
$$

with $g^{(mj)} = g^{(mj)}\cdot e^{i\cdot\varphi_{\Omega(mj)}};\quad i = \sqrt{-1};$

$$\tag{8}$$

As this structure looks a bit complicated in the case of general dependent "quantum" or wave function vectors f_m, we refer to the examples given by Schwarzer [6, 7] for better orientation. In addition to the elaborations in Ref. 6, we should point out here that the equations are invariant under phase transformation, meaning any of the resulting (tuples) equations will remain the same under transformations of the form:

$$\bar{g}^{(mj)} = \tilde{g}^{(mj)} \cdot e^{i \cdot \phi} = g^{(mj)} \cdot e^{i \cdot \varphi_{\Omega(mj)}} \cdot e^{i \cdot \phi}; \quad i = \sqrt{-1}; \quad \varphi_{\Omega(mj)} = \{0, \pi\}. \quad (9)$$

The same also holds for all the other tuple terms under the square root as long as, after squaring the roots, a reasonable parameter for the angle has been chosen. Here is one example:

$$\text{vec} = \left\{ \begin{array}{l} \sqrt{(\alpha \exp[i \cdot \phi 2] + \beta \exp[i \cdot (\phi 1 + \pi)] + \epsilon \exp[i \cdot (\phi + 0)])^2} \\ \sqrt{(\alpha \exp[i \cdot \phi 2] + \beta \exp[i \cdot (\phi 1 + \pi)] + \epsilon \exp[i \cdot (\phi + \pi)])^2} \\ \sqrt{(\alpha \exp[i \cdot \phi 2] + \beta \exp[i \cdot (\phi 1 + 0)] + \epsilon \exp[i \cdot (\phi + \pi/2)])^2} \\ \sqrt{(\alpha \exp[i \cdot \phi 2] + \beta \exp[i \cdot (\phi 1 + 0)] + \epsilon \exp[i \cdot (\phi - \pi/2)])^2} \end{array} \right\}$$

$$\text{with} \quad \text{vec} \cdot \text{vec} = 4 \left(\alpha^2 \exp[2 \cdot i \cdot \phi 2] + \beta^2 \exp[2 \cdot i \cdot \phi 1] \right) \quad (10)$$

where we obtain the desired $\text{vec} \cdot \text{vec} = 4 \left(\alpha^2 + \beta^2 \right)$ for all sets $\phi 1 = \phi 2 = n \cdot \pi/2$ with $n = 0, \pm 1, \pm 2, \ldots$

For more about the 4-vector concept by Stamler and Co., we must refer to Ref. 2.

In Refs. 6 and 7, Schwarzer gave a rather harsh criticism about the work of the third author in Ref. 2, who also happens to be the main author of this work. It read:

It is not clear to the present author why the third author in [2], who apparently put the story together on his own, because the two Stamlers had long been dead (killed in Auschwitz) the moment the publication [2] was written, did not see (or simply considered it too obvious to explicitly mention it) that the most natural parameter to apply this substructure approach to, should be the line element of a general metric. With this metric fulfilling the Einstein field equations, a quantized GTR should be the outcome. In this case, the whole pretty complicated looking mathematical structure in [2] becomes relatively simple.

Bodan accepted this criticism with a generous "Point well taken!" (c.f. [8]). He also suggested the following (still [8]):

> Why did Schwarzer not see (or simply considered it too obvious to explicitly mention it) that the most natural explanation why his vector or V_Ω approach works or has to be applied at all, lies in the fact that obviously all its components behave like coordinates?
>
> Why can we treat the components of the V_Ω (or vec as it was named above) like coordinates?
>
> The answer, as it often is the case when being in the comfortable position of the relaxed reader watching from outside rather than the author sitting in the thick of things, seems to be obvious. We can treat the V_Ω-components like independent coordinates because they are respectively they have become independent coordinates. Depending on the processes within space creates additional properties or axes. They evolve as derivatives from the existing axes, becoming apparently independent when only seen (being observed) from a certain scale. Thus, the processes in space do not only shape the space-time regarding its metric but also lead to the evolution of degrees of freedom, which is to say to new dimensions.

The reader should bear that in mind when following the Schwarzer approach in quantizing the line element as we are simply going to repeat it in the following section. We will also hit the "additional dimension aspect," meaning the "production" of additional dimensions by certain states and processes in space-time again when discussing the photon.

Chapter 2

Theory

The Generalized Metric Dirac Operator

Contrary to the Stamler approach [2] and own our earlier approaches [6, 7] where the starting point was the Einstein–Hilbert action, and thus the general theory of relativity, we here start with an adapted Dirac ansatz.

It is easy to prove that the scalar product of the vector

$$V_\Omega = \begin{Bmatrix} a+b+c+d+e, a+b+c+d-e, a+b+c-d+e, a+b+c-d-e, \\ a+b-c+d+e, a+b-c+d-e, a+b-c-d+e, a+b-c-d-e, \\ a-b+c+d+e, a-b+c+d-e, a-b+c-d+e, a-b+c-d-e, \\ a-b-c+d+e, a-b-c+d-e, a-b-c-d+e, a-b-c-d-e \end{Bmatrix}$$
$$\equiv [a \pm b \pm c \pm d \pm e]_\Omega \qquad (11)$$

is

$$V_\Omega \cdot V_\Omega = a^2 + b^2 + c^2 + d^2 + e^2. \qquad (12)$$

Even the introduction of "virtual" parameters ε_i can be incorporated into the vector V_Ω as follows:

$$V_\Omega = \{a+b+c, a+b-c, a-b+i\cdot c, a-b-i\cdot c\} \equiv [a \pm b \pm I \cdot c]_\Omega$$
$$\text{with} \quad V_\Omega \cdot V_\Omega = a^2 + b^2$$
$$(13)$$

The Theory of Everything: Quantum and Relativity is Everywhere – A Fermat Universe
Norbert Schwarzer
Copyright © 2020 Jenny Stanford Publishing Pte. Ltd.
ISBN 978-981-4774-47-5 (Hardcover), 978-1-315-09975-0 (eBook)
www.jennystanford.com

with c now playing the role of such a virtual parameter. Now we introduce the vector V_Ω:

$$V_\Omega = \left[g^{m0} \frac{\partial f_m}{\partial y^0} \pm g^{m1} \frac{\partial f_m}{\partial y^1} \pm g^{m2} \frac{\partial f_m}{\partial y^2} \pm g^{m3} \frac{\partial f_m}{\partial y^3} \pm I \cdot \varepsilon^m \cdot f_m \right.$$
$$\left. \pm i \cdot \sqrt{\tilde{\lambda}^m} \cdot p_m \cdot f_m \right]_\Omega$$
$$\equiv \left[g^{m0} \frac{\partial f_m}{\partial y^0} \pm g^{m1} \frac{\partial f_m}{\partial y^1} \pm g^{m2} \frac{\partial f_m}{\partial y^2} \pm g^{m3} \frac{\partial f_m}{\partial y^3} \pm I \cdot \varepsilon^m \cdot f_m \right.$$
$$\left. \pm i \cdot \sqrt{\lambda^m} \cdot f_m \right]_\Omega . \tag{14}$$

Here the m-index summation has to be understood as

$$\sqrt{\lambda^m} \cdot f_m =$$
$$\left\{ \begin{array}{l} \sqrt{\lambda^0} \cdot f_0 + \sqrt{\lambda^1} \cdot f_1 + \sqrt{\lambda^2} \cdot f_2 + \sqrt{\lambda^3} \cdot f_3, \ \sqrt{\lambda^0} \cdot f_0 + \sqrt{\lambda^1} \cdot f_1 + \sqrt{\lambda^2} \cdot f_2 - \sqrt{\lambda^3} \cdot f_3, \\ \sqrt{\lambda^0} \cdot f_0 + \sqrt{\lambda^1} \cdot f_1 - \sqrt{\lambda^2} \cdot f_2 + \sqrt{\lambda^3} \cdot f_3, \ \sqrt{\lambda^0} \cdot f_0 + \sqrt{\lambda^1} \cdot f_1 - \sqrt{\lambda^2} \cdot f_2 - \sqrt{\lambda^3} \cdot f_3, \\ \sqrt{\lambda^0} \cdot f_0 - \sqrt{\lambda^1} \cdot f_1 + \sqrt{\lambda^2} \cdot f_2 + \sqrt{\lambda^3} \cdot f_3, \ \sqrt{\lambda^0} \cdot f_0 - \sqrt{\lambda^1} \cdot f_1 + \sqrt{\lambda^2} \cdot f_2 - \sqrt{\lambda^3} \cdot f_3, \\ \sqrt{\lambda^0} \cdot f_0 - \sqrt{\lambda^1} \cdot f_1 - \sqrt{\lambda^2} \cdot f_2 + \sqrt{\lambda^3} \cdot f_3, \ \sqrt{\lambda^0} \cdot f_0 - \sqrt{\lambda^1} \cdot f_1 - \sqrt{\lambda^2} \cdot f_2 - \sqrt{\lambda^3} \cdot f_3 \end{array} \right\} . \tag{15}$$

Thus, the total number of vector components counted by the index Ω depends on the number of summands of our vector expression. For the vector given in Eq. (14), we find the following scalar product:

$$V_\Omega \cdot V_\Omega = g^{mi} \frac{\partial f_m}{\partial y^i} \cdot g^{nj} \frac{\partial f_n}{\partial y^j} - \sqrt{\tilde{\lambda}^m} \cdot p_m \cdot f_m \cdot \sqrt{\tilde{\lambda}^n} \cdot p_n \cdot f_n. \tag{16}$$

Denoting the eigenvectors of the metric g_{ij} with e_i (e^i for g^{ij}) and considering the $\tilde{\lambda}^m$ as the eigenvalues of the contravariant metric tensor g^{ij}, we can evaluate

$$V_\Omega \cdot V_\Omega = g^{mi} \frac{\partial f_m}{\partial y^i} \cdot g^{nj} \frac{\partial f_n}{\partial y^j} - \sqrt{\tilde{\lambda}^m} \cdot p_m \cdot f_m \cdot \sqrt{\tilde{\lambda}^n} \cdot p_n \cdot f_n$$
$$= g^{mi} \frac{\partial f}{\partial y^i} e_m g^{nj} \frac{\partial f}{\partial y^j} e_n - \sqrt{\lambda^m} e_m \cdot \sqrt{\lambda^n} e_n$$
$$= g^{mi} \frac{\partial x^\alpha}{\partial y^i} \sqrt{\tilde{\lambda}_\alpha} \tilde{e}_\alpha g^{nj} \frac{\partial x^\beta}{\partial y^j} \sqrt{\tilde{\lambda}_\beta} \tilde{e}_\beta e_m e_n - g^{nm} e_n e_m$$
$$= g^{mi} \frac{\partial x^\alpha}{\partial y^i} g^{nj} \frac{\partial x^\beta}{\partial y^j} \tilde{g}_{\alpha\beta} e_m e_n - g^{nm} e_n e_m$$
$$= g^{mi} \frac{\partial f}{\partial y^i} e_m g^{nj} \frac{\partial f}{\partial y^j} e_n - g^{nm} e_n e_m$$
$$\Rightarrow V_\Omega \cdot V_\Omega = g^{mi} g^{nj} g_{ij} e_m e_n - g^{nm} e_n e_m = g^{mn} e_m e_n - g^{nm} e_n e_m. \tag{17}$$

Considering the first and second term after the first "=" sign separately,

$$g^{mi}\frac{\partial f_m}{\partial y^i} \cdot g^{nj}\frac{\partial f_n}{\partial y^j} = g^{mi}\frac{\partial f}{\partial y^i}e_m g^{nj}\frac{\partial f}{\partial y^j}e_n = g^{mi}\frac{\partial f}{\partial y^i}g^{nj}\frac{\partial f}{\partial y^j}e_m e_n$$

$$\sqrt{\tilde{\lambda}^m} \cdot p_m \cdot f_m \cdot \sqrt{\tilde{\lambda}^n} \cdot p_n \cdot f_n = \sqrt{\lambda^m}f_m \cdot \sqrt{\lambda^n}f_n = \sqrt{g^{mn}} \cdot e_n \cdot \sqrt{g^{nm}} \cdot e_m$$

$$= g^{nm}e_m e_n, \tag{18}$$

we see that we have used the spectral decomposition of a symmetric tensor and the tensor transformation rule. The rest is defining the properties of the function f—that is, the function vector f_m. Now we multiply both terms of Eq. (18), in other words, both sides of the last equation in Eq. (17), with an infinitesimal scalar element dy and define the infinitesimal vector element $dy_m = e_m*dy$. This gives

$$V_\Omega \cdot V_\Omega \cdot dy^2 = g^{mn}e_m e_n \cdot dy^2 - g^{nm}e_n e_m \cdot dy^2$$

$$= g^{mn}dy_m dy_n - g^{nm}dy_n dy_m$$

$$= ds^2 - ds^2 = 0$$

$$\Rightarrow V_\Omega \cdot V_\Omega = g^{mi}\frac{\partial f_m}{\partial y^i} \cdot g^{nj}\frac{\partial f_n}{\partial y^j} - \sqrt{\tilde{\lambda}^m} \cdot p_m \cdot f_m \cdot \sqrt{\tilde{\lambda}^n} \cdot p_n \cdot f_n = 0.$$

$$\tag{19}$$

For simplicity, we demand the above equation to be zero by constructing zero-value factors, or vector components, in the following way:

$$\Rightarrow V_\Omega = \left\{ i \cdot \sqrt{\tilde{\lambda}^m} \cdot p_m \cdot f_m \pm g^{mi}\frac{\partial f_m}{\partial y^i} \right\}_\Omega$$

$$= \begin{cases} i \cdot \sqrt{\tilde{\lambda}^m} \cdot p_m \cdot f_m + g^{mi}\frac{\partial f_m}{\partial y^i} \\ i \cdot \sqrt{\tilde{\lambda}^m} \cdot p_m \cdot f_m - g^{mi}\frac{\partial f_m}{\partial y^i} \end{cases}_\Omega = 0. \tag{20}$$

The corresponding tuples, or vector components, are already demonstrated above. Due to the similarity with the Dirac approach [5], we might call the operator $\widehat{V}_\Omega^{\,m}$, defined as

$$\Rightarrow \widehat{V}_\Omega^{\,m} f_m = \left\{ g^{mi}\frac{\partial}{\partial y^i} \pm i \cdot \sqrt{\tilde{\lambda}^m} \cdot p_m \right\}_\Omega f_m \tag{21}$$

the generalized metric Dirac operator.

One might say that this operator demands that in essence everything must either change with respect to space and time and then the thing be proportional to its changes or it simply does not, in face cannot, be observable in our scale. As the introduction of "virtual" parameters has shown, there can be much more "in the back ground" of the other scales, but it does not appear (directly) observable to us within our scale (only indirectly via other observables).

Scalar Product in Laplace–Beltrami Form

We will now try to obtain the scalar product of our tuple vectors in such a way that the complete Laplace operators will be reproduced. Building the scalar product by applying an operator of the form $\frac{1}{\sqrt{g}}\left\{\widehat{V}_\Omega^{\overset{m}{}}\right\}\sqrt{g}$ from the left and multiplying with another f vector from the right to our generalized metric Dirac operator (21) leaves us with

$$
0 = \frac{1}{\sqrt{g}}\left\{\widehat{V}_\Omega^{\overset{n}{}}\right\}\sqrt{g} \cdot \left\{\widehat{V}_\Omega^{\overset{m}{}}\right\} f_m f_n
$$

$$
= \Bigg[-\left(\sqrt{\lambda^0}\right)^2 - \left(\sqrt{\lambda^\Sigma}\right)^2 + g^{00}\underbrace{\frac{1}{\sqrt{g}}\partial_0\sqrt{g}\cdot g^{00}\partial_0}_{\Delta-\text{Operator-term}}
$$

$$
+ g^{\Sigma\Sigma}\underbrace{\frac{1}{\sqrt{g}}\partial_\Sigma\sqrt{g}\cdot g^{\Sigma\Sigma}\partial_\Sigma}_{\Delta-\text{Operator-term}} \Bigg]\Psi
$$

$$
f_n f_m \rightarrow \Psi; \quad g = \det\left(g_{ij}\right); \quad \Delta = 4D - \text{Laplace Operator}
$$

$$(22)$$

where we recognize the four-dimensional Laplace operator terms for our completely arbitrary metric. In Minkowski coordinates the Laplace or Laplace–Beltrami operator is also called D'Alembert operator and is given with the sign $W \equiv \partial^\mu \partial_\mu = \frac{\partial_t^2}{c^2} - \Delta_{\text{Cartesian}}; \Delta_{\text{Cartesian}} = \partial_x^2 + \partial_y^2 + \partial_z^2$. However, here we will use Δ as the general symbol for the Laplace operator in arbitrary coordinates, or in an arbitrary metric. There we always have

$$
\Delta_{\text{Metric}} = \sum_{i,j=1}^{D}\frac{1}{\sqrt{g}}\partial_i\sqrt{g}\cdot g^{ij}\partial_j; \quad D = \text{Dimension.} \quad (23)
$$

In order to obtain a complete and clean Laplace operator in Eq. (22), an additional factor g_{nm} has to be applied to the left-hand side of our scalar product. With the short form

$$
0 = \left[i\sqrt{\lambda^m}f_m + \left(g^{mi}\partial_i\right)f_m\right]_{[\Omega]} = \left[\left(i\sqrt{\lambda^m} + g^{mi}\partial_i\right)f_m\right]_{[\Omega]} \equiv \widehat{V}_\Omega^{\overset{m}{}}f_m
$$

$$
\widehat{V}_\Omega^{\overset{m}{}} = \left[\left\{i\sqrt{\lambda^m}\right\} + \left\{g^{mi}\partial_i\right\}, \left\{i\sqrt{\lambda^m}\right\} - \left\{g^{mi}\partial_i\right\}\right] \quad (24)
$$

we can write

$$0 = g_{nm} \frac{1}{\sqrt{g}} \left\{ \overset{n}{\hat{V}}_\Omega \right\} \sqrt{g} \cdot \left\{ \overset{m}{\hat{V}}_\Omega \right\} f_m f_n$$

$$= g_{nm} \frac{1}{\sqrt{g}} \begin{bmatrix} \left\{ i\sqrt{\lambda^n} \right\} + \left\{ g^{nm} \partial_m \right\}, \\ \left\{ i\sqrt{\lambda^n} \right\} - \left\{ g^{nm} \partial_m \right\} \end{bmatrix} \sqrt{g} \cdot \begin{bmatrix} \left\{ i\sqrt{\lambda^m} \right\} + \left\{ g^{mi} \partial_i \right\}, \\ \left\{ i\sqrt{\lambda^m} \right\} - \left\{ g^{mi} \partial_i \right\} \end{bmatrix} f_m f_n$$

$$= \frac{1}{\sqrt{g}} \begin{bmatrix} \left\{ i\sqrt{\lambda_m} \right\} + \left\{ \delta^n_m \partial_m \right\}, \\ \left\{ i\sqrt{\lambda_m} \right\} - \left\{ \delta^n_m \partial_m \right\} \end{bmatrix} \sqrt{g} \cdot \begin{bmatrix} \left\{ i\sqrt{\lambda^m} \right\} + \left\{ g^{mi} \partial_i \right\}, \\ \left\{ i\sqrt{\lambda^m} \right\} - \left\{ g^{mi} \partial_i \right\} \end{bmatrix} f_m f_n$$

with $g_{nm} g^{nm} = \delta^m_m = 4; \quad g_{nm} \sqrt{\lambda^n} = g_{nm} \sqrt{\tilde{\lambda}^n} p_n = \sqrt{\tilde{\lambda}_m} p_m$ (25)

with $\sqrt{\tilde{\lambda}_m} p_m$ standing for $\sqrt{g_{mm}} p_m$. Please note that for simplicity so far we have not considered the V_Ω vector structure for non-diagonal metrics and refer to Refs. 9–11 and later in this book with respect to such metrics also containing shear components. Building the scalar product of all tuples gives

$$0 = g_{nm} \frac{1}{\sqrt{g}} \left\{ \overset{n}{\hat{V}}_\Omega \right\} \sqrt{g} \cdot \left\{ \overset{m}{\hat{V}}_\Omega \right\} f_m f_n$$

$$= g_{nm} \begin{bmatrix} \underbrace{\left(\sqrt{g^{mn}} \cdot \sqrt{g^{mn}} \right) p^2_m}_{\begin{array}{l} = g_{nm} \sqrt{\tilde{\lambda}^n} \cdot \sqrt{\tilde{\lambda}^m} \cdot p^2_m \\ = \sqrt{\tilde{\lambda}_m \tilde{\lambda}^m} \cdot p^2_m = p^2_m \sqrt{4} \end{array}} + \underbrace{\frac{g^{mn}}{\sqrt{g}} \partial_m \sqrt{g} \cdot g^{mi} \partial_i}_{\Delta - \mathrm{Operator}} \end{bmatrix} f_n f_m$$

$$= \begin{bmatrix} \underbrace{g_{nm} \left(\sqrt{g^{mn}} \cdot \sqrt{g^{mn}} \right) p^2_m}_{\begin{array}{l} = g_{nm} \left(\sqrt{g^{mn}} \cdot \sqrt{g^{mn}} \right) p^2_m \\ = \delta^n_n \cdot p^2_m = 4 \cdot p^2_m \end{array}} + \frac{g_{nm} g^{mn}}{\sqrt{g}} \partial_m \sqrt{g} \cdot g^{mi} \partial_i \end{bmatrix} f_n f_m$$

$$f_n f_m \to \Psi; \quad g = \det\left(g_{ij} \right); \quad \Delta = 4D - \text{Laplace Operator} \quad (26)$$

which simplifies to

$$0 = g_{nm}\frac{1}{\sqrt{g}}\left\{\widehat{V}_{\Omega}^{n}\right\}\sqrt{g}\cdot\left\{\widehat{V}_{\Omega}^{m}\right\}f_m f_n$$

$$= 4\cdot\left[p_m^2 + \underbrace{\frac{1}{\sqrt{g}}\partial_m\sqrt{g}\cdot g^{mi}\partial_i}_{\Delta-\text{Operator}}\right]f_n f_m$$

$$f_n f_m \to \Psi; \quad g = \det\left(g_{ij}\right); \quad \Delta_{\text{Metric}} = 4D - \text{Laplace Operator}$$

$$\Rightarrow \quad \boxed{0 = \left[p_m^2 + \Delta_{Metric}\right]\Psi}. \tag{27}$$

This surprisingly simple result is in almost (see further below) perfect agreement with the required condition that in the limit to the Minkowski space the equation above shall reproduce the Klein–Gordon equation, which is given as

$$\left[\frac{\partial_t^2}{c^2} - \Delta_{\text{Cartesian}} + \frac{M^2 c^2}{\hbar^2}\right]\Psi = 0. \tag{28}$$

On the other side, rewriting Eq. (27) in Minkowski coordinates leads to

$$0 = g_{nm}\frac{1}{\sqrt{g}}\left\{\widehat{V}_{\Omega}^{n}\right\}\sqrt{g}\cdot\left\{\widehat{V}_{\Omega}^{m}\right\}f_m f_n = 4\cdot\left[p_m^2 + \underbrace{\frac{1}{\sqrt{g}}\partial_m\sqrt{g}\cdot g^{mi}\partial_i}_{\Delta_{\text{Metric}}-\text{Operator}}\right]f_n f_m$$

$$\Rightarrow \quad 0 = \left[p_m^2 + \left(\frac{\partial_t^2}{c^2} - \Delta_{\text{Cartesian}}\right)\right]\Psi . \tag{29}$$

Comparison with Eq. (28) gives

$$p_m^2 = \frac{M^2 c^2}{\hbar^2} \quad \Rightarrow \quad \sqrt{p_m^2} = \frac{M\cdot c}{\hbar}. \tag{30}$$

So far, when considering this outcome separately from our other Dirac-like solutions (c.f. [6–17]), it did not matter where the additional metric factor g_{nm}, introduced in Eq. (26), would come from. Especially as the result for p_m was equivalent to what was obtained directly from the Dirac equation. So, for instance, when quantizing the Schwarzschild metric [17] via the generalized Dirac operator, we obtain the solution (cf. section "Quantized Schwarzschild

Metric" further below) for a Schwarzschild particle at rest as follows:

$$f_{(0)}(t, r) = \{C_{01} \cdot g_1(t), C_{02} \cdot g_1(t), C_{03} \cdot g_2(t), C_{04} \cdot g_2(t)\}$$

$$\lim_{r \to \infty} g_1 = \lim_{r \to \infty} e^{-\frac{i \cdot c^2 \left(e^{\frac{i \cdot j \cdot \pi}{p}} E_P + \mu \sqrt{\frac{r}{c^2(r-r_s)}} \right) \cdot (r-r_s) \cdot t}{r}}$$

$$= e^{-i \cdot c^2 t \left(e^{\frac{i \cdot j \cdot \pi}{p}} E_P + \frac{\mu}{c} \right)} \xrightarrow{E_P = 0} e^{-i \cdot c \cdot t \cdot \mu}$$

$$\lim_{r \to \infty} g_2 = \lim_{r \to \infty} e^{\frac{i \cdot c^2 \left(e^{\frac{i \cdot j \cdot \pi}{p}} E_P + \mu \sqrt{\frac{r}{c^2(r-r_s)}} \right) \cdot (r-r_s) \cdot t}{r}}$$

$$= e^{i \cdot c^2 t \left(e^{\frac{i \cdot j \cdot \pi}{p}} E_P + \frac{\mu}{c} \right)} \xrightarrow{E_P = 0} e^{i \cdot c \cdot t \cdot \mu}$$

with $\quad \mu = p_m$. $\hspace{4cm}$ (31)

For here and now it is not important how this solution has to be understood (for this we refer to the section "Quantized Schwarzschild Metric"), because we are only interested in the limit of the radius r to infinity. This, together with $E_P = 0$, should give us the classical Dirac solutions for the particle at rest [5], and we are able to determine the parameter $\mu = p_m$ as

$$\mu = p_m = \frac{c \cdot M}{\hbar}. \hspace{3cm} (32)$$

Comparing with the outcome of Eq. (30) we see that there is perfect agreement, but we should still ask how the so far "artificial" introduction of g_{mn} could be omitted without changing the result in the limit to the classical cases. In order to explain this here and now, we have to apply the identity $\mathbf{f}_n \cdot \mathbf{f}_m = \mathbf{e}_n \cdot \mathbf{e}_m h^2 = g_{nm} h^2$, which will be elaborated in section "Further Considerations." This allows us to rewrite Eq. (27) in the following form:

$$0 = \frac{1}{\sqrt{g}} \left\{ \hat{V}_\Omega^{\,n} \right\} \sqrt{g} \cdot \left\{ \hat{V}_\Omega^{\,m} \right\} f_m f_n$$

$$= \left[g^{mn} \cdot p_m^2 + \underbrace{\frac{g^{nm}}{\sqrt{g}} \partial_m \sqrt{g} \cdot g^{mi} \partial_i}_{\Delta-\text{Operator}} \right] f_n f_m$$

$$= \left[g^{mn} \cdot p_m^2 + g^{nm} \underbrace{\frac{1}{\sqrt{g}} \partial_m \sqrt{g} \cdot g^{mi} \partial_i}_{\Delta-\text{Operator}} \right] g_{nm} h^2$$

$$= g_{nm} \left[g^{mn} \cdot p_m^2 + \frac{g^{nm}}{\sqrt{g}} \partial_m \sqrt{g} \cdot g^{mi} \partial_i \right] g_{nm} h^2 + h^2 \frac{g^{nm}}{\sqrt{g}} \partial_m \sqrt{g} \cdot g^{mi} \partial_i g_{nm}$$

$$= 4 \cdot \left[p_m^2 + \frac{1}{\sqrt{g}} \partial_m \sqrt{g} \cdot g^{mi} \partial_i \right] h^2 + h^2 \cdot \frac{g^{nm}}{\sqrt{g}} \partial_m \sqrt{g} \cdot g^{mi} \partial_i g_{nm}$$

$$\Psi \equiv h^2 \Rightarrow \boxed{0 = \left[p_m^2 + \Delta_{\text{Metric}} \right] \Psi + \frac{\Psi}{4} \cdot \frac{g^{nm}}{\sqrt{g}} \partial_m \sqrt{g} \cdot g^{mi} \partial_i g_{nm}} \ . \qquad (33)$$

Still, and as expected, we obtain the correct limits in the case of the Minkow-ski metric, but we also recognize the additional term $\frac{\Psi}{4} \cdot \frac{g^{nm}}{\sqrt{g}} \partial_m \sqrt{g} \cdot g^{mi} \partial_i g_{nm}$ in comparison with the classical Klein–Gordon equation. For a flat space with constant metric coefficients, this term gives zero, because we would have to derive

$$\partial_m \sqrt{g} \cdot g^{mi} \partial_i \begin{bmatrix} \Downarrow \\ g_{nm} \end{bmatrix} = \partial_m \begin{bmatrix} \Downarrow \\ \sqrt{g} \cdot g^{mi} \end{bmatrix} \cdot \partial_i \begin{bmatrix} \Downarrow \\ g_{nm} \end{bmatrix} + \sqrt{g} \cdot g^{mi} \partial_m \partial_i \begin{bmatrix} \Downarrow\Downarrow \\ g_{nm} \end{bmatrix} . \qquad (34)$$

For non-constant metrics, however, our new Klein–Gordon equation obtains additional derivatives in dependence on the metric. It must be emphasized that this result is questionable, because it contradicts the idea of the Ω vectors, whose intention it was to result in independent separable operators and but which led to funny results (see the example of the Schwarzschild metric and Eq. (37) below). Thus, we intend to formally resolve that problem by reconsidering the derivation (33) in the following manner:

$$E_n h^2 = \frac{1}{\sqrt{g}} \left\{ \widehat{V}_\Omega^{\,n} \right\} \sqrt{g} \cdot \left\{ \widehat{V}_\Omega^{\,m} \right\} f_m f_n = \left[g^{mn} \cdot p_m^2 + \underbrace{\frac{g^{nm}}{\sqrt{g}} \partial_m \sqrt{g} \cdot g^{mi} \partial_i}_{\Delta-\text{Operator}} \right] \mathbf{f}_n \cdot \mathbf{f}_m$$

$$g_{nm} E_n h^2 g^{mn} = g_{nm} \left[g^{mn} \cdot p_m^2 + \underbrace{g^{nm} \frac{1}{\sqrt{g}} \partial_m \sqrt{g} \cdot g^{mi} \partial_i}_{\Delta-\text{Operator}} \right] g_{nm} h^2 g^{mn}$$

$$4 \cdot E_n h^2 = 4^2 \cdot \left[p_m^2 + \frac{1}{\sqrt{g}} \partial_m \sqrt{g} \cdot g^{mi} \partial_i \right] h^2$$

$$\Psi \equiv h^2 \Rightarrow \boxed{0 = \left[p_m^2 + \Delta_{\text{Metric}} \right] \Psi - \frac{E_n}{4} \cdot \Psi} \ . \qquad (35)$$

Please note the simple multiplication with metric tensors from left and right. Here, for the reason of later generalization, we had already considered possible eigenvalues E_n. This equation is much simpler. Nevertheless, in the following, for completeness, we will here also give the version (33).

Still we have the metric of the space-time preserved in our new Klein–Gordon equation. Therefore, we might also call the new equation the metric Klein–Gordon equation.

It should also be noted that only with the two equations, namely the metric Klein–Gordon (35) and the metric Dirac equation (e.g., in the form (20)), the solution for both the quantum metric g_{ij} and the scalar quantum solution h can be found and properly separated from each other.

Example: Schwarzschild Metric

As the Schwarzschild metric, for historic as well as scientific reasons, is one of the most important Einstein field solutions, we will give the complete generalized Klein–Gordon equation for this metric. While the metric Dirac equation will extensively be treated later in this book, for the Schwarzschild metric we here want to consider its "metric square" only for completeness without further discussion. Using the original Schwarzschild coordinates and metric as given in Eq. (120) (in contravariant form), we obtain the following metric Klein–Gordon equation:

$$0 = \left[p_m^2 + \Delta_{\text{Schwarzschild}} \right] \Psi + \frac{\Psi}{4} \cdot \frac{g^{nm}}{\sqrt{g}} \partial_m \sqrt{g} \cdot g^{mi} \partial_i g_{nm}$$

$$0 = \left[\frac{r(r_s - r)\partial_r^2 + (r_s - 2r)\partial_r - \csc[\theta]^2 \partial_\phi^2 - \partial_\theta^2 - \cot[\theta]\partial_\theta}{r^2} + \frac{r}{c^2(r - r_s)} \partial_t^2 + p_m^2 \right] \Psi \ . \tag{36}$$

For completeness we also give the version after Eq. (33):

$$0 = \left[p_m^2 + \Delta_{\text{Schwarzschild}} \right] \Psi + \frac{\Psi}{4} \cdot \frac{g^{nm}}{\sqrt{g}} \partial_m \sqrt{g} \cdot g^{mi} \partial_i g_{nm}$$

$$\Rightarrow 0 = \left[\frac{r(r_s - r)\partial_r^2 + (r_s - 2r)\partial_r - \csc[\theta]^2 \partial_\phi^2 - \partial_\theta^2 - \cot[\theta]\partial_\theta}{r^2} \right.$$

$$\left. + \frac{r}{c^2(r - r_s)} \partial_t^2 \right] \Psi$$

$$+ \left[\frac{\left(5r_s + r \left(-2 + \frac{r}{-r + r_s} \right) - 2r \csc[\theta]^2 \right)}{2 \cdot r^3} + p_m^2 \right] \Psi; \quad \Psi \equiv h^2.$$

$$\tag{37}$$

Knowing that the Schwarzschild radius $r_S = \frac{2 \cdot G \cdot m}{c^2}$ is connected with the mass m and having seen that in the limit to the flat Minkowski space for both the metric Dirac and the metric Klein–Gordon equation our parameter p_m must result in $p_m = \frac{c \cdot M}{\hbar}*$ (see Eq. (32)), we might suspect to have a quantization condition for the mass of a Schwarzschild object being connected with our Eq. (37).

It should be pointed out that Eq. (37) has to be considered the Klein–Gordon or quantum equation for a Schwarzschild object.

Here, however, we do not intend to solve this equation, but leave this task for later in another publication.

Transition to the Metric Schrödinger or Covariant Schrödinger Equation

Because for many people *the* quantum equation has been seen in the Schrödinger equation, our next task in this book will be to give a most general way to obtain the Schrödinger equation out of our metric Klein–Gordon Eq. (35). This time we do not intend to obtain the classical Schrödinger equation, but its covariant analogy.

For simplicity, we assume a metric with none of the components actually being time dependent. The component g^{00} shall be of the form $g^{00} = C_1/c^2$, C_1 being a constant, and all other metric time components shall be $g^{0i} = 0$ ($i = 1, 2, 3$). As it is simple to erase the metric derivative term later, we start with the questionable Klein–Gordon form (33) (last line). Then we can separate the time derivative as follows:

$$0 = \left[p_m^2 + \Delta_{\text{Metric}} \right] \Psi + \frac{\Psi}{4} \cdot g^{nm} \Delta_{\text{Metric}} g_{nm}$$

$$= \left[p_m^2 + \underbrace{\frac{1}{\sqrt{g}} \partial_m \sqrt{g} \cdot g^{mi} \partial_i}_{\Delta-\text{Operator}} \right] \Psi + \frac{\Psi}{4} \cdot \frac{g^{nm}}{\sqrt{g}} \partial_m \sqrt{g} \cdot g^{mi} \partial_i g_{nm}$$

$$= \left[p_m^2 + \left(C_1 \cdot \frac{\partial_t^2}{c^2} + \underbrace{\frac{1}{\sqrt{g}} \partial_\alpha \sqrt{g} \cdot g^{\alpha\beta} \partial_\beta}_{3D-\Delta-\text{Operator}} \right) \right] \Psi + \frac{\Psi}{4} \cdot g^{nm} \left(C_1 \cdot \frac{\partial_t^2}{c^2} + \underbrace{\frac{1}{\sqrt{g}} \partial_\alpha \sqrt{g} \cdot g^{\alpha\beta} \partial_\beta}_{3D-\Delta-\text{Operator}} \right) g_{nm}$$

$$= \left[p_m^2 + \left(C_1 \cdot \frac{\partial_t^2}{c^2} + \frac{1}{\sqrt{g}} \partial_\alpha \sqrt{g} \cdot g^{\alpha\beta} \partial_\beta \right) \right] \Psi + \frac{\Psi}{4} \cdot g^{nm} \left(\frac{1}{\sqrt{g}} \partial_\alpha \sqrt{g} \cdot g^{\alpha\beta} \partial_\beta \right) g_{nm}. \quad (38)$$

---

*The distinction between the symbols M and m for the parameter mass in Einstein's field theory (m) and quantum theory (M) will become clear at some point later when we consider the matter–antimatter asymmetry.

In order to later come to the consistent form (35) (last line), we simply have to erase all terms where the operator $\frac{g^{nm}}{\sqrt{g}}\partial_m\sqrt{g}\cdot g^{mi}\partial_i$ acts on the metric only (cf. final result as given in). Now we introduce the function $\Psi = \Phi + X$ and demand the following additional condition:

$$\partial_t\Psi = c^2 \cdot C_2 \cdot (\Phi - X).\tag{39}$$

Together with (38) we obtain

$$0 = \left[p_m^2(\Phi+X) + \left(C_1\cdot C_2\cdot\partial_t(\Phi-X) + \frac{1}{\sqrt{g}}\partial_\alpha\sqrt{g}\cdot g^{\alpha\beta}\partial_\beta(\Phi+X)\right)\right]$$
$$+\frac{\Phi+X}{4}\cdot g^{nm}\left(\frac{1}{\sqrt{g}}\partial_\alpha\sqrt{g}\cdot g^{\alpha\beta}\partial_\beta\right)g_{nm}.\tag{40}$$

The following two equations summed up would result in (40)

$$0 = \left[\Phi\left\{p_m^2 + g^{nm}\left(\frac{1}{\sqrt{g}}\partial_\alpha\sqrt{g}\cdot g^{\alpha\beta}\partial_\beta\right)g_{nm}\right\}\right.$$
$$\left.+\left(C_1\cdot C_2\cdot\partial_t\Phi + \frac{1}{2\sqrt{g}}\partial_\alpha\sqrt{g}\cdot g^{\alpha\beta}\partial_\beta(\Phi+X)\right)\right]$$

$$0 = \left[X\left\{p_m^2 + g^{nm}\left(\frac{1}{\sqrt{g}}\partial_\alpha\sqrt{g}\cdot g^{\alpha\beta}\partial_\beta\right)g_{nm}\right\}\right.$$
$$\left.+\left(-C_1\cdot C_2\cdot\partial_t X + \frac{1}{2\sqrt{g}}\partial_\alpha\sqrt{g}\cdot g^{\alpha\beta}\partial_\beta(\Phi+X)\right)\right].\tag{41}$$

Comparing with the original Schrödinger equation as given in the form below

$$\left[-i\cdot\hbar\cdot\partial_t - \frac{\hbar^2}{2M}\Delta_{\text{Schrödinger}} + V\right]\Psi = 0\tag{42}$$

does not only give us the matter and antimatter solutions again (cf. [6–17] and various sections of this book, e.g., "Quantized Schwarzschild Metric") but also shows us—as discussed before [12]—that the potential V has now been taken on by the metric and the proportionally factors p_m. Thus, a distorted metric acts like an effective potential in the Schrödinger approximation and vice versa, which is to say, what in classical physics has been described as a potential would now become a distorted metric providing the necessary interaction. Disregarding the antimatter solution here, the constants C_1 and C_2 and the proportionally factors can be obtained as

$$C_1\cdot C_2 = -\frac{i\cdot M}{\hbar};\quad X = 0$$

$$\Rightarrow\left[-\frac{1}{2}\Delta_{\text{Schrödinger}} + V\right]\Psi = \left[\begin{array}{l}\Phi\left\{p_m^2 + g^{nm}\left(\frac{1}{\sqrt{g}}\partial_\alpha\sqrt{g}\cdot g^{\alpha\beta}\partial_\beta\right)g_{nm}\right\}\\[2mm]+\left(\frac{1}{2\sqrt{g}}\partial_\alpha\sqrt{g}\cdot g^{\alpha\beta}\partial_\beta\Phi\right)\end{array}\right].$$
$$\tag{43}$$

Now we erase the metric derivative term and obtain

$$C_1 \cdot C_2 = -\frac{i \cdot m}{\hbar}; \quad X = 0$$

$$\Rightarrow \left[-\tfrac{1}{2} \Delta_{\text{Schrödinger}} + V \right] \Psi = \left[\left(\mu_m^2 + \frac{1}{2\sqrt{g}} \partial_\alpha \sqrt{g} \cdot g^{\alpha\beta} \partial_\beta \right) \Phi \right], \tag{44}$$

where we have the desired metric Schrödinger equation:

$$0 = \left[\left(\mu_m^2 + \frac{1}{2\sqrt{g}} \partial_\alpha \sqrt{g} \cdot g^{\alpha\beta} \partial_\beta \right) \Phi \right]. \tag{45}$$

Things get a bit more interesting when it is assumed that V was chosen such that we have $\Psi = \Psi_{\text{Schrödinger}} = \Phi_{\text{new}} = \Phi$. Then we can directly write down the equation for the determination of the potential and its connection to the metric distortion:

$$V = \mu_m^2 + \frac{1}{2\Psi} \left(\frac{1}{\sqrt{g}} \partial_\alpha \sqrt{g} \cdot g^{\alpha\beta} \partial_\beta - \Delta_{\text{Flat\_Space}} \right) \Psi \tag{46}$$

where we have applied the sign convention of GTR (general theory of relativity) with the spatial components having negative signs ($\Delta_{\text{Schrödinger}} = -\Delta_{\text{Flat\_Space}}$). When also using the result from (32), we can write

$$V = \frac{m^2 c^2}{\hbar^2} + \frac{1}{2\Psi} \left(\frac{1}{\sqrt{g}} \partial_\alpha \sqrt{g} \cdot g^{\alpha\beta} \partial_\beta - \Delta_{\text{Flat\_Space}} \right) \Psi \tag{47}$$

which, principally, also gives us the means to derive a metric to a given potential. However, as we have seen in the previous paper [12] that mass itself seems to be a dynamic parameter, or a parameter arising from a dynamic, but apparently a stationary process (stationary for us, that is, on our scale), it appears wise to just leave the proportionality factors p_m unfixed for now. They will probably show their character in connection with the task described by Eq. (46) the moment the potential V is known.

We note: what classically is the potential V gives a metric distortion in the theory of everything.

Further Considerations

The general tools as outlined above now need to be applied to line elements in such a way that we are able to distinguish between the classical element and its quantum part. So, we could easily interpret the "clever zero" in Eq. (19) as one being obtained for the quantum part of the line element. Thus, we can write

$$\frac{ds^2_{\text{quantum}} - ds^2_{\text{quantum}}}{dy^2} = (g_{ij} + q_{ij})\, \epsilon^i \epsilon^j - (g^{nm} + q^{nm})\, \epsilon_n \epsilon_m \tag{48}$$

$$\text{with} \quad ds^2_{\text{total}} = \left(ds_{\text{classic}} + ds_{\text{quantum}} \right)^2$$

where we apply the symbol $\in_m$ for the quantum base vector and q^{ij} for the quantum metric, with the total line element being given as

$$
\begin{aligned}
ds_{\text{total}}^2 &= \gamma^{nm} d\xi_n d\xi_m = \gamma^{nm} (dx_n + dy_n)(dx_m + dy_m) \\
&= (g^{nm} + q^{nm})(e_n dx + \in_n dy)(e_m dx + \in_m dy) \\
&= \gamma^{nm} dx_n dx_m + \gamma^{nm} dx_n dy_m + \gamma^{nm} dy_n dx_m + \gamma^{nm} dy_n dy_m \\
&= \gamma^{nm}(dx_n dx_m + dx_n d(p_m f_m) + d(p_n f_n) dx_m + d(p_n f_n) d(p_m f_m)).
\end{aligned}
$$

$$(49)$$

Now we assume that $\gamma^{ij} \simeq g^{ij}$ is motivated by considering g^{ij} a base metric and q^{ij} its small perturbation (cf. [15]) and find

$$
\begin{aligned}
ds_{\text{total}}^2 - ds_{\text{total}}^2 &= \gamma^{nm} d\xi_n d\xi_m - \gamma_{nm} d\xi^n d\xi^m \\
&\simeq g^{nm}(dx_n dx_m + dx_n d(p_m f_m) + d(p_n f_n) dx_m \\
&\quad + d(p_n f_n) d(p_m f_m)) - g_{nm}(dx^n dx^m + dy^n dx^m + dx^n dy^m \\
&\quad + dy^n dy^m) \\
&= g^{nm}(dx_n dx_m + dx_n d(p_m f_m) + d(p_n f_n) dx_m \\
&\quad + d(p_n f_n) d(p_m f_m)) - g_{nm}\left(dx^n dx^m + d\left(\gamma^{ni}\frac{\partial f}{\partial x^i}\right) dx^m \right. \\
&\quad \left. + dx^n d\left(\gamma^{mi}\frac{\partial f}{\partial x^i}\right) + d\left(\gamma^{ni}\frac{\partial f}{\partial x^i}\right) d\left(\gamma^{mi}\frac{\partial f}{\partial x^i}\right)\right) \\
&\simeq g^{nm}(dx_n dx_m + dx_n d(p_m f_m) + d(p_n f_n) dx_m \\
&\quad + d(p_n f_n) d(p_m f_m)) - g_{nm}\left(dx^n dx^m + d\left(g^{ni}\frac{\partial f}{\partial x^i}\right) dx^m \right. \\
&\quad \left. + dx^n d\left(g^{mi}\frac{\partial f}{\partial x^i}\right) + d\left(g^{ni}\frac{\partial f}{\partial x^i}\right) d\left(g^{mi}\frac{\partial f}{\partial x^i}\right)\right). \quad (50)
\end{aligned}
$$

Now we can easily apply our recipe from above to all terms containing dy_k, or dy^k, and obtain

$$
\begin{aligned}
\Rightarrow \overset{\text{total}}{d\xi_n} &= \overset{\text{classic}}{dx_n} + \overset{\text{quantum}}{dy_n} = \overset{\text{classic}}{dx_n} + \overset{\text{quantum}}{d(p_n f_n(x))} \\
\Rightarrow \overset{\text{total}}{\xi_n} &= \overset{\text{classic}}{x_n} + \overset{\text{quantum}}{y_n} = \overset{\text{classic}}{x_n} + \overset{\text{quantum}}{p_n f_n(x)}
\end{aligned}
$$

$$(51)$$

with the corresponding solution for the functions $f_k(x)$ as it results from our extended (metric) Dirac equation (20).

Please note that the approximation $\gamma^{ij} \simeq g^{ij}$ is not a must in the approach above. We could easily assume to know the combined metric for the quantized theory and still perform the same evaluation as above without any approximation or additional assumption. Then we would obtain

$$ds_{total}^2 - ds_{total}^2 = 0 = \gamma^{nm} d\xi_n d\xi_m - \gamma_{nm} d\xi^n d\xi^m$$

$$= \gamma^{nm} (dx_n dx_m + d\mathbf{x}_n h \cdot \mathbf{e}_m dy + h \cdot \mathbf{e}_n dy \cdot d\mathbf{x}_m$$

$$+ \mathbf{e}_n \cdot \mathbf{e}_m h^2 dy^2) - \gamma_{nm} \left(dx^n dx^m + d \left(\gamma^{ni} \frac{\partial h}{\partial x^i} \right) dx^m \right.$$

$$\left. + dx^n d \left(\gamma^{mi} \frac{\partial h}{\partial x^i} \right) + d \left(\gamma^{ni} \frac{\partial h}{\partial x^i} \right) d \left(\gamma^{mj} \frac{\partial h}{\partial x^j} \right) \right)$$

$$= \gamma^{nm} (dx_n dx_m + d\mathbf{x}_n \cdot \mathbf{f}_m dy + \mathbf{f}_n dy \cdot d\mathbf{x}_m + \mathbf{f}_n \cdot \mathbf{f}_m dy^2)$$

$$- \left(\gamma_{nm} dx^n dx^m + d \left(\gamma^{ni} \frac{\partial \mathbf{f}_n}{\partial x^i} \right) \cdot \mathbf{e}_m dx^m \right.$$

$$\left. + d \left(\gamma^{mi} \frac{\partial \mathbf{f}_m}{\partial x^i} \right) \cdot \mathbf{e}_n dx^n + d \left(\gamma^{ni} \frac{\partial \mathbf{f}_n}{\partial x^i} \right) d \left(\gamma^{mj} \frac{\partial \mathbf{f}_m}{\partial x^j} \right) \right)$$

$$\Rightarrow \overset{\text{total}}{d\xi_n} = \overset{\text{classic}}{dx_n} + \overset{\text{quantum}}{dy_n} = \overset{\text{classic}}{dx_n} + \overset{\text{quantum}}{d (\mathbf{e}_n h (x))}$$

$$\Rightarrow \overset{\text{total}}{\xi_n} = \overset{\text{classic}}{\mathbf{x}_n} + \overset{\text{quantum}}{\mathbf{y}_n} = \overset{\text{classic}}{\mathbf{x}_n} + \overset{\text{quantum}}{\mathbf{f}_n (x)} \tag{52}$$

where, in many cases, such as diagonal metrics, this simplifies to

$$ds_{total}^2 - ds_{total}^2 = \gamma^{nm} d\xi_n d\xi_m - \gamma_{nm} d\xi^n d\xi^m$$

$$= \gamma^{nm} (dx_n dx_m + dx_n d (p_m f_m) + d (p_n f_n) dx_m + d (p_n f_n) d (p_m f_m))$$

$$- \gamma_{nm} (dx^n dx^m + dy^n dx^m + dx^n dy^m + dy^n dy^m)$$

$$= \gamma^{nm} (dx_n dx_m + dx_n d (p_m f_m) + d (p_n f_n) dx_m + d (p_n f_n) d (p_m f_m)) - \gamma_{nm}$$

$$\left(dx^n dx^m + d \left(\gamma^{ni} \frac{\partial f}{\partial x^i} \right) dx^m + dx^n d \left(\gamma^{mi} \frac{\partial f}{\partial x^i} \right) + d \left(\gamma^{ni} \frac{\partial f}{\partial x^i} \right) d \left(\gamma^{mi} \frac{\partial f}{\partial x^i} \right) \right)$$

$$\Rightarrow \overset{\text{total}}{d\xi_n} = \overset{\text{classic}}{dx_n} + \overset{\text{quantum}}{dy_n} = \overset{\text{classic}}{dx_n} + \overset{\text{quantum}}{d (p_n f_n (x))}$$

$$\Rightarrow \overset{\text{total}}{\xi_n} = \overset{\text{classic}}{\mathbf{x}_n} + \overset{\text{quantum}}{y_n} = \overset{\text{classic}}{\mathbf{x}_n} + \overset{\text{quantum}}{p_n f_n (x)} . \tag{53}$$

A comprehensive list of generalizations of the method can be found in Ref. 9.

Example: The Classical Dirac Equation in the Minkowski Space-Time and Its Extension to Arbitrary Coordinates

More or less in order to give a demonstration without the need of referring to external literature and in order to show where to introduce the scale parameter, we repeat here one of our earlier considerations regarding the Dirac equation and its classical solutions.

At first, we investigate under which conditions the equations derived above provide the solutions as they are obtained from the Dirac equation [5]. As a brief reminder, we shall recall that the Dirac equation based upon the Dirac operator D_i is defined as

$$D_i = \left(A \cdot \hat{p}_x + B \cdot \hat{p}_y + C \cdot \hat{p}_z + D \cdot \frac{i}{c} \frac{\partial}{\partial t} \right)$$
$$= i\hbar \left(A \cdot \partial_x + B \cdot \partial_y + C \cdot \partial_z + D \cdot \partial_t \right). \tag{54}$$

It is being used in the Dirac equation

$$\left(D_i - m \cdot c^2 \right) \Psi = 0 \tag{55}$$

with the Dirac matrices

$$\gamma^1 = A = \begin{pmatrix} 0 & 0 & 0 & 1 \\ 0 & 0 & 1 & 0 \\ 0 & -1 & 0 & 0 \\ -1 & 0 & 0 & 0 \end{pmatrix}; \quad \gamma^2 = B = \begin{pmatrix} 0 & 0 & 0 & -i \\ 0 & 0 & i & 0 \\ 0 & i & 0 & 0 \\ -i & 0 & 0 & 0 \end{pmatrix},$$

$$\gamma^3 = C = \begin{pmatrix} 0 & 0 & 1 & 0 \\ 0 & 0 & 0 & -1 \\ -1 & 0 & 0 & 0 \\ 0 & 1 & 0 & 0 \end{pmatrix}; \quad \gamma^0 = D = \begin{pmatrix} 1 & 0 & 0 & 0 \\ 0 & 1 & 0 & 0 \\ 0 & 0 & -1 & 0 \\ 0 & 0 & 0 & -1 \end{pmatrix}. \tag{56}$$

These give four solutions for a particle (fermion) at rest, namely for two particles representing ordinary matter and the corresponding antiparticle with two possible spins for each.

Besides, using the Dirac notation for the gamma matrices, we can write the covariant form of the Dirac equation:

$$D_i = i \cdot \gamma^\mu \partial_\mu \quad \Rightarrow \quad \left(i \cdot \gamma^\mu \partial_\mu - \frac{m \cdot c^2}{\hbar} \right) \Psi = 0. \tag{57}$$

Here we will only consider the particle at rest solution:

$$\Psi(t)_r = \left\{ \begin{pmatrix} e^{-i \cdot t \cdot M} \\ 0 \\ 0 \\ 0 \end{pmatrix}, \begin{pmatrix} 0 \\ e^{-i \cdot t \cdot M} \\ 0 \\ 0 \end{pmatrix}, \begin{pmatrix} 0 \\ 0 \\ e^{i \cdot t \cdot M} \\ 0 \end{pmatrix}, \begin{pmatrix} 0 \\ 0 \\ 0 \\ e^{i \cdot t \cdot M} \end{pmatrix} \right\}. \tag{58}$$

with $M = \frac{m \cdot c^2}{\hbar}$

Now we apply our approach and start with the line element vector [7]:

$$ds_\Omega = \left(\sqrt{\lambda^m} \cdot f_m\right)_\Omega \cdot dy$$

$$\Rightarrow 0 = \left[\sqrt{\left(g^{m0}\frac{\partial f_m}{\partial y^0} \pm g^{m1}\frac{\partial f_m}{\partial y^1} \pm g^{m2}\frac{\partial f_m}{\partial y^2} \pm g^{m3}\frac{\partial f_m}{\partial y^3} \pm I \cdot \varepsilon^m \cdot f_m \pm i \cdot \sqrt{\lambda^m} \cdot f_m\right)^2}\right]_\Omega .$$

$$(59)$$

It should be noted that in the case of metric shear the form (59) has to be adapted [9–11], but as we are only interested in a principle comparison with the Dirac solution for a particle at rest in connection with an additional free parameter (here the metric shear component), we choose here a simplified treatment of the problem. After having introduced the V_Ω forms also in the metric shear cases, we will see how the evaluation has to be performed in an unapproximated manner (cf. section "The Photon").

For simplicity, we set

$$m^m = \sqrt{\lambda^m} .$$

$$(60)$$

Also setting $\varepsilon = 0$, assuming a metric with one shear component g^{30} (like in the Kerr metric [18, 19]) and "particle at rest," and thus an f vector depending only on time, we obtain the following differential equations:

$$0 = \left[\sqrt{\left(g^{00}\partial_0 f_0 + g^{30}\partial_0 f_3 \pm i \cdot \sqrt{|g|} \cdot m^m \cdot f_m\right)^2}\right]_{[\Omega]}$$

$$= \frac{m^m f_{m[\Omega]}}{\sqrt{-\left(m^m f_{m[\Omega]}\right)^2}} \left[-m^m f_m + i \left(g^{00}\partial_0 f_0 + g^{30}\partial_0 f_3\right)\right]_{[\Omega]}$$

with $\quad g^{(mj)} = g^{(mj)} \cdot e^{i \cdot \varphi_{(mj)}}; \quad i = \sqrt{-1}; \quad \varphi_{(00)} = \{0, \pi\}; \quad \varphi_{(30)} = \{0, \pi\}$

$$\Rightarrow 0 = \begin{bmatrix} -m^0 f_0 + i \cdot \left|g^{00}\right| \partial_0 f_0 \\ -m^3 f_3 + i \cdot \left|g^{30}\right| \partial_0 f_3 \\ -m^0 f_0 - i \cdot \left|g^{00}\right| \partial_0 f_0 \\ -m^3 f_3 - i \cdot \left|g^{30}\right| \partial_0 f_3 \end{bmatrix} .$$

$$(61)$$

The solutions read

$$\begin{pmatrix} f_{(0)}(t) \\ f_{(3)}(t) \end{pmatrix} = \left\{ \begin{pmatrix} C_{01} \cdot e^{-\frac{i \cdot t \cdot m^{(0)}}{|g^{00}|}} \\ C_{31} \cdot e^{-\frac{i \cdot t \cdot m^{(3)}}{|g^{30}|}} \end{pmatrix}, \begin{pmatrix} C_{02} \cdot e^{-\frac{i \cdot t \cdot m^{(0)}}{|g^{00}|}} \\ C_{32} \cdot e^{\frac{i \cdot t \cdot m^{(3)}}{|g^{30}|}} \end{pmatrix}, \begin{pmatrix} C_{03} \cdot e^{\frac{i \cdot t \cdot m^{(0)}}{|g^{00}|}} \\ C_{33} \cdot e^{\frac{i \cdot t \cdot m^{(3)}}{|g^{30}|}} \end{pmatrix}, \right.$$

$$\left. \begin{pmatrix} C_{04} \cdot e^{\frac{i \cdot t \cdot m^{(0)}}{|g^{00}|}} \\ C_{34} \cdot e^{-\frac{i \cdot t \cdot m^{(3)}}{|g^{30}|}} \end{pmatrix} \right\} .$$

$$(62)$$

We see that to each matter $-\frac{i \cdot t \cdot m^{(0)}}{|g^{00}|}$ and antimatter $+\frac{i \cdot t \cdot m^{(0)}}{|g^{00}|}$ solution belong two solutions with exponents $\pm\frac{i \cdot t \cdot m^{(3)}}{|g^{30}|}$. Even though this is only an approximation, it can easily be shown that this solution mathematically handles the Pauli exclusion [20] without the need of a Clifford or a spinor algebra as necessary for the Dirac solutions. For more about this solution, the reader is referred to Ref. 7.

The only aspect we are interested in here is the question where we have to introduce the scaling or quantum uncertainty parameter usually connected with the Planck constant $\hbar$. For this, we simply need to compare the Dirac solution for the particle at rest with our solution given in Eq. (62) for f_0.

$$M = \frac{m_{\text{Dirac}} \cdot c^2}{\hbar} \Leftrightarrow \frac{m^{(0)}}{\left|g^{00}\right|} = \frac{\sqrt{\lambda^{(0)}}}{\left|g^{00}\right|}. \tag{63}$$

Knowing that the eigenvalues of the metric are usually proportional to the metric components (e.g., in the Kerr case), we obtain*

$$\lambda^i = \begin{pmatrix} g^{11} \\ g^{22} \\ \frac{1}{2}\left(g^{00} + g^{33} - \sqrt{\left(g^{00}\right)^2 + 4\left(g^{30}\right)^2 - 2g^{00}g^{33} + \left(g^{33}\right)^2}\right) \\ \frac{1}{2}\left(g^{00} + g^{33} + \sqrt{\left(g^{00}\right)^2 + 4\left(g^{30}\right)^2 - 2g^{00}g^{33} + \left(g^{33}\right)^2}\right) \end{pmatrix}. \tag{64}$$

We can easily extract from Eq. (63) that the phase extension of the metric as applied in Eq. (61) (or principally in Eq. (8)) should be written as

$$g^{(mj)} \rightarrow \quad \sim g^{(mj)} \cdot \hbar^2 \cdot e^{i \cdot \varphi_{(mj)}}; \quad i = \sqrt{-1}. \tag{65}$$

Thus, the square of the scale parameter determining the principle uncertainty of a system has to be multiplied with the tensor component one intends to quantize. In the case of a metric, this can immediately be understood, because the metric components are proportional to the squared line element ds^2, automatically making ds directly proportional to the uncertainty parameter (which in quantum mechanics happens to be just $\hbar$).

---

*By handling the metric shear components correctly with respect to the V_Ω vector and also by taking their complex coordinate dependence into account, Schwarzer was meanwhile able to derive a complete solution for the Kerr particle at rest [10]. It is a revolving wave, rotating around the Kerr axis of symmetry. As we here only want to introduce the new general uncertainty parameter $\hbar$, it is of no importance which example we choose and how correctly we handle the shear components or whether or not we take their coordinate dependence into account. For more, the reader is referred to Ref. 10 and the section "Quantization of the Kerr Metric."

When Max Planck introduced his famous constant h (or $\hbar = h/2\,\pi$), he used this letter because it stood for the word "help." As we, too, want to apply a help parameter assisting us in controlling the uncertainties of a certain tensor field (or a whole theory), we want to keep this letter. Thus, for further use we will now introduce a generalized uncertainty parameter $\not{\hslash}$, which will be applied as before in Eq. (65), but now being of a much more general, which is to say tensorial, character. It shall be defined in connection with an arbitrary tensor T^{ij} as

$$T^{(\alpha\beta)} \rightarrow \quad T^{(\alpha\beta)} \cdot \left(\not{\hslash}^{(\alpha\beta)}\right)^2 \cdot e^{i\cdot\varphi_{(\alpha\beta)}}; \quad i = \sqrt{-1}. \tag{66}$$

With this, regarding the application to any scientific field, things are very straightforward now.

We simply have to determine the principle uncertainty for every tensor field influencing (or potentially influencing) our mathematical model, or mathematical model space, and "phase-adapt" all tensor components as shown above by multiplying them with the uncertainty parameter squared. The subsequent linear differential equations of first order will give us the probability functions for our model. Other applications and alternative approaches can be found in Refs. 8, 11–13.

The Connection to the Einstein Field Equations

Already having introduced the metric tensor into our theory, we can easily assure the conformity to the Einstein field theory by demanding the metric to fulfill the Einstein field equations [3, 4]:

$$R^{\alpha\beta} - \frac{1}{2}Rg^{\alpha\beta} + \Lambda g^{\alpha\beta} = -\kappa T^{\alpha\beta} \Rightarrow \frac{R^{\alpha\beta} + \kappa T^{\alpha\beta}}{\frac{1}{2}R - \Lambda} = g^{\alpha\beta}. \tag{67}$$

In the non-approximated general case of Eq. (52), we demand the relativistic and quantum conform metric γ^{ij} to fulfill the Einstein field equation, which gives

$$^{\gamma}R^{\alpha\beta} - \frac{1}{2}{}^{\gamma}R\gamma^{\alpha\beta} + \Lambda\gamma^{\alpha\beta} = -\kappa T^{\alpha\beta} \Rightarrow \frac{^{\gamma}R^{\alpha\beta} + \kappa T^{\alpha\beta}}{\frac{1}{2}{}^{\gamma}R - \Lambda} = \gamma^{\alpha\beta} \tag{68}$$

with the new symbols $^{\gamma}R^{\alpha\beta}$ and $^{\gamma}R$ denoting the Ricci tensor and the Ricci scalar being evaluated from the metric γ^{ij}, respectively.

Thus, simply by setting the metric tensor and its eigenvalues in accordance with Eq. (1), or Eq. (68), we will have a consistent general quantum theory of relativity.

The well-informed reader will immediately see that there is no reason to restrict the technique applied in Eqs. (6) and (7) to the metric tensor of an Einstein–Hilbert action. Any tensor, and in those cases where we will need the spectral decomposition of a tensor, any symmetric tensor can be quantized this way, and this is what we have made use of within applications apparently being far away from the quantum theory of relativity. So, it was shown that the method can also be applied to tackle the principle uncertainty problem in material science [13], socioeconomy and finance [8].

Summing Up the Recipe: The Forward Derivation

We intend to find the quantum part of a metric γ^{ij}, which is to say we search for the quantum field connected with this metric.

(1) We separate the coordinates of this metric $\xi_n = \overset{\text{total}}{\xi_n}$ into a classical and a quantum related part: $\overset{\text{total}}{\xi_n} = \overset{\text{classic}}{x_n} + \overset{\text{quantum}}{y_n} = \overset{\text{classic}}{x_n} + F\,(f_n\,(x)) \cdot f_n\,(x)$. Here F simply stands for a function of the function vector f_m, and in many cases this is just a linear connector like

$$K\,(f_n\,(x)) = F\,(f_n\,(x)) \cdot f_n\,(x) = p_n \cdot f_n\,(x)$$
$$= \{p_0 \cdot f_0,\ p_1 \cdot f_1,\ p_2 \cdot f_2,\ p_3 \cdot f_3\}.$$

This gives the linear coordinate elements as

$$\overset{\text{total}}{d\xi_n} = \overset{\text{classic}}{dx_n} + \overset{\text{quantum}}{dy_n} = \overset{\text{classic}}{dx_n} + dK\,(f_n\,(x)).$$

(2) We introduce these elements into the "clever-zero" equation of two total line elements as $ds_{\text{total}}^2 - ds_{\text{total}}^2 = \gamma^{nm}d\xi_n d\xi_m - \gamma_{nm}d\xi^n d\xi^m$ and substitute accordingly for the covariant elements $d\xi_n$ plus set a gradient expression for their contravariant brothers like

$$d\xi^n = dx^n + \gamma^{ni}\frac{\partial f}{\partial x^i}, \quad d\xi^m = dx^m + \gamma^{mi}\frac{\partial f}{\partial x^i}.$$

(3) Forming the square root in the Ω form given in Eq. (20) results in a distinct number (depending on the number of independent metric components) of generalized Dirac-like equations for the functions f_m, given as

$$\widehat{V}_\Omega^{\,m} f_m = \left\{ g^{mi}\frac{\partial}{\partial y^i} \pm i \cdot \sqrt{g^{mi}} \cdot e_i \cdot \hat{F} \right\}_\Omega f_m. \tag{69}$$

(4) Forming the scalar product of the Ω forms in the form $g_{nm} \frac{1}{\sqrt{g}} \left\{ \overset{n}{\hat{V}}_\Omega \right\} \sqrt{g} \cdot$ $\left\{ \overset{m}{\hat{V}}_\Omega \right\} f_m$ gives us a metric Klein–Gordon-like equation for the scalar core h^2 of the f_m vectors (with the definition $\mathbf{f}_n \cdot \mathbf{f}_m = \mathbf{e}_n \cdot \mathbf{e}_m h^2 = g_{nm} h^2 \equiv \Psi \cdot g_{nm}$):

$$ 0 = \left[-\hat{F}^2 + \square_{\text{Metric}} \right] \Psi + \frac{\Psi}{4} \cdot \frac{g^{nm}}{\sqrt{g}} \partial_m \sqrt{g} \cdot g^{mi} \partial_i g_{nm}. \tag{70} $$

(5) Thereby, the functional operator $\hat{F}$ can be extracted from the energy momentum tensor being defined in the symmetric Hilbert form: $T^{\alpha\beta} = \frac{2}{\sqrt{-g}} \frac{\delta \left(\sqrt{-g} \cdot L_M \right)}{\delta g_{\alpha\beta}}$ and the Euler–Lagrange equations for L_M, which should be the Lagrange density for matter.

Remark: It should be noted for generality that it would also be possible to demand a more complex functional approach for the gradient setting within the V_Ω vectors. Thus, one might like to consider cases like the following one:

$$ \overset{m}{\hat{V}}_\Omega f_m = \left\{ D^m \pm i \cdot \sqrt{g^{mi}} \cdot e_i \cdot \hat{F} \right\}_\Omega f_m $$

$$ = \left\{ D_1 \cdot g^{mi} \frac{\partial}{\partial y^i} \pm D_2 \cdot e^m g^{ij} \frac{\partial}{\partial y^i} \frac{\partial}{\partial y^j} \pm \ldots \pm i \cdot \sqrt{g^{mi}} \cdot e_i \cdot \hat{F} \right\}_\Omega f_m. $$

$$ \tag{71} $$

Even though within the scope of usual physics such an extension does not seem to be of need, we will still seek for possible extension later in this book (cf. section "Generalization of 'The Recipe': From $\hbar$ to the Planck Tensor").

Summing Up the Recipe: The Backward Derivation

This time we start with a given energy–momentum tensor and intend to find the quantum part of the metric γ^{ij}, which is to say we search for the metric connected with this tensor. Thereby we assume the quantum field of the matter being completely given due to the energy–momentum tensor.

(1) We set $\gamma^{ij} = g^{ij}$ and extract the functional operator $\hat{F}$ from the energy–momentum tensor being defined in the symmetric Hilbert form: $T^{\alpha\beta} = \frac{2}{\sqrt{-g}} \frac{\delta \left(\sqrt{-g} \cdot L_M \right)}{\delta g_{\alpha\beta}}$. Thereby, the Euler–Lagrange equations for L_M, which is the Lagrange density for matter, will always take on the form $0 = \left[-\hat{F}^2 + \square_{\text{Metric}} \right] \Psi + \frac{\Psi}{4} \cdot \frac{g^{nm}}{\sqrt{g}} \partial_m \sqrt{g} \cdot g^{mi} \partial_i g_{nm}$. Thus, it gives us the desired functional operator $\hat{F}$.

(2) The next step is to apply the new concept of forming the square root in the Ω form described in the theory section by assuming $g_{nm} \frac{1}{\sqrt{g}} \left\{ \hat{V}_\Omega^n \right\} \sqrt{g} \cdot \left\{ \hat{V}_\Omega^m \right\} \mathbf{f}_n \cdot \mathbf{f}_m = 0 = \left[-\hat{F}^2 + \Box_{\text{Metric}} \right] \Psi + \frac{\Psi}{4} \cdot \frac{g^{nm}}{\sqrt{g}} \partial_m \sqrt{g} \cdot g^{mi} \partial_i g_{nm}$. This results in an Ω number of Dirac-like equations (refraining from writing $\mathbf{f}_m$ in bold further on):

$$\hat{V}_\Omega^m f_m = \left\{ g^{mi} \frac{\partial}{\partial y^i} \pm i \cdot \sqrt{g^{mi}} \cdot e_i \cdot \hat{F} \right\}_\Omega f_m = 0. \tag{72}$$

(3) After solving these equations and considering the coordinates $\overset{\text{total}}{\xi_n} = \overset{\text{total}}{\xi_n}$ of the unknown metric γ^{ij} to be separable as follows: $\overset{\text{total}}{\xi_n} = \overset{\text{classic}}{x_n} + \overset{\text{quantum}}{y_n} = \overset{\text{classic}}{x_n} + \underbrace{F\left(f_n\left(x\right)\right) \cdot f_n\left(x\right)}_{\equiv K(f_n(x))}$, we found the quantum part of the deformable coordinates to be $\overset{\text{quantum}}{y_n} = K\left(f_n\left(x\right)\right)$, which—this way—has originally been defined by the matter part of the Einstein–Hilbert action.

Example: The Higgs Field

We choose the Lagrange density L_M as a function of a scalar parameter w like

$$L_M = \gamma^{ij} w_{,i} w_{,j} - \mu^2 w^2 + \lambda^2 w^4. \tag{73}$$

The next step is to build the Euler–Lagrange equations, which gives us

$$\Box_{\text{Metric}} w + \mu^2 w - \lambda^2 w^3 \equiv \Box_{\text{Metric}} w - F^2 \left[w \right] \equiv \left[\Box_{\text{Metric}} - \hat{F}^2 \right] w = 0 \tag{74}$$

and where we recognize the Klein–Gordon equation for the setting $\lambda = 0$.

Comparison with Eq. (70) would give us $w = f_m$, but as w, as per definition, must be a scalar function we have to assume that the operator $\left[\Box_{\text{Metric}} - \hat{F}^2 \right]$ only acts on one f_m component at a time. We assume that the contribution of the term $\frac{\Psi}{4} \cdot \frac{g^{nm}}{\sqrt{g}} \partial_m \sqrt{g} \cdot g^{mi} \partial_i g_{nm}$ can be ignored. It would only change the metric to be applied within further evaluation. As we keep it general for now, no harm is done in not considering the Laplacian of the metric.

The corresponding generalized Dirac equations are already given in Ref. 15 and read

$$\Rightarrow \hat{V}_\Omega^m f_m = \left\{ i \cdot \sqrt{\tilde{\lambda}^m} \cdot F\left[f_m \right] f_m \pm g^{mi} \frac{\partial f_m}{\partial y^i} \right\}_\Omega$$

$$= \left\{ i \cdot \sqrt{\tilde{\lambda}^m} \cdot \left(\mu \pm i \cdot \lambda \cdot f_m \right) f_m \pm g^{mi} \frac{\partial f_m}{\partial y^i} \right\}_\Omega = 0 \tag{75}$$

where we have sucked all factors into the constants. One easily recognizes the self-interaction in connection with the non-linear term $f_m^* f_m$ in the term $(\mu \pm i \cdot \lambda \cdot f_m) f_m$.

After separation the subsequent equations regarding the 0 components read

$$\Rightarrow \widehat{V}^0_\Omega w = \left\{ i \cdot \sqrt{g^{0j}} \cdot e_j \cdot (\mu \pm i \cdot \lambda \cdot w) \pm g^{0i} \frac{\partial}{\partial x^i} \right\}_\Omega w = 0, \quad (76)$$

where we have assumed that the parameter w is only connected with the metric components $m = 0$. Please note that our choice of the index $m = 0$ is completely arbitrary and has, for the time being, nothing to do with the usual setting $x_0 = t = \text{time}$. Further assuming only diagonal metrics, we obtain

$$\widehat{V}^0_\Omega w\left(x^0\right) = \left\{ i \cdot \sqrt{g^{00}} \cdot \left(\mu \pm i \cdot \lambda \cdot w\left(x^0\right)\right) \pm g^{00} \frac{\partial}{\partial x^0} \right\}_\Omega w\left(x^0\right) = 0 \quad (77)$$

which results in the following differential equations:

$$\begin{cases} i \cdot \sqrt{g^{00}} \cdot \left(\mu + i \cdot \lambda \cdot w\left(x^0\right)\right) + g^{00} \frac{\partial}{\partial x^0} \\ i \cdot \sqrt{g^{00}} \cdot \left(\mu - i \cdot \lambda \cdot w\left(x^0\right)\right) + g^{00} \frac{\partial}{\partial x^0} \\ i \cdot \sqrt{g^{00}} \cdot \left(\mu + i \cdot \lambda \cdot w\left(x^0\right)\right) - g^{00} \frac{\partial}{\partial x^0} \\ i \cdot \sqrt{g^{00}} \cdot \left(\mu - i \cdot \lambda \cdot w\left(x^0\right)\right) - g^{00} \frac{\partial}{\partial x^0} \end{cases} w\left(x^0\right) = 0 \quad (78)$$

giving the results ($C[k]$ just being constants)

$$w\left(x^0\right) = \begin{cases} \pm e^{i_\mu \left(\frac{x^0}{\sqrt{g^{00}}} + C[1]\right)} \frac{\mu}{-i\lambda} \\ \pm e^{i_\mu \left(\frac{x^0}{\sqrt{g^{00}}} - C[2]\right)} \frac{\mu}{+i\lambda} \\ \pm e^{-i_\mu \left(\frac{x^0}{\sqrt{g^{00}}} - C[3]\right)} \frac{\mu}{-i\lambda} \\ \pm e^{-i_\mu \left(\frac{x^0}{\sqrt{g^{00}}} + C[4]\right)} \frac{\mu}{+i\lambda} \end{cases}_\Omega. \quad (79)$$

Taking now $w = \overset{\text{total}}{\xi_0} = \overset{\text{classic}}{x_0} + \overset{\text{quantum}}{y_0} = \overset{\text{classic}}{x_0} + K(f_0(x \overset{\text{quantum}}{=} x_0 = g_{00} x^0))$ we find the quantum distortion of the spatial coordinate x_0 caused by the Higgs field to be

$$
\overset{total}{\xi_0} = x_0 + K\left(w\left(x^0\right)\right) = x_0 + \left\{ \begin{array}{l} K\left(\dfrac{\mu}{\pm e^{i_\mu\left(\frac{x^0}{\sqrt{g^{00}}}+C[1]\right)}-i\lambda}\right) \\[2em] K\left(\dfrac{\mu}{\pm e^{i_\mu\left(\frac{x^0}{\sqrt{g^{00}}}-C[2]\right)}+i\lambda}\right) \\[2em] K\left(\dfrac{\mu}{\pm e^{-i_\mu\left(\frac{x^0}{\sqrt{g^{00}}}-C[3]\right)}-i\lambda}\right) \\[2em] K\left(\dfrac{\mu}{\pm e^{-i_\mu\left(\frac{x^0}{\sqrt{g^{00}}}+C[4]\right)}+i\lambda}\right) \end{array} \right\}_\Omega , \tag{80}
$$

$$
\overset{total}{\xi_0} = x_0 + \left\{ \begin{array}{l} \mu\left(\dfrac{\mu}{\pm e^{i_\mu\left(\frac{x^0}{\sqrt{g^{00}}}+C[1]\right)}-i\lambda}\right)+i\lambda\left(\dfrac{\mu}{\pm e^{i_\mu\left(\frac{x^0}{\sqrt{g^{00}}}+C[1]\right)}-i\lambda}\right)^2 \\[2em] \mu\left(\dfrac{\mu}{\pm e^{i_\mu\left(\frac{x^0}{\sqrt{g^{00}}}-C[2]\right)}+i\lambda}\right)-i\lambda\left(\dfrac{\mu}{\pm e^{i_\mu\left(\frac{x^0}{\sqrt{g^{00}}}-C[2]\right)}+i\lambda}\right)^2 \\[2em] \mu\left(\dfrac{\mu}{\pm e^{-i_\mu\left(\frac{x^0}{\sqrt{g^{00}}}-C[3]\right)}-i\lambda}\right)-i\lambda\left(\dfrac{\mu}{\pm e^{-i_\mu\left(\frac{x^0}{\sqrt{g^{00}}}-C[3]\right)}-i\lambda}\right)^2 \\[2em] \mu\left(\dfrac{\mu}{\pm e^{-i_\mu\left(\frac{x^0}{\sqrt{g^{00}}}+C[4]\right)}+i\lambda}\right)+i\lambda\left(\dfrac{\mu}{\pm e^{-i_\mu\left(\frac{x^0}{\sqrt{g^{00}}}+C[4]\right)}+i\lambda}\right)^2 \end{array} \right\}_\Omega .
$$
$$\tag{81}$$

Thus, we have obtained the spatial "quantum wiggle" causing the Higgs field for the coordinate x_0.

In the same way as outlined above, Higgs–eigenvalue problems could be considered and allow the discussion of particle masses. This, however, will be discussed later in this book (see also [9]).

Example: Eigenvalue Solutions for Simple Fields with $K(f_m) = F(f_m)^* f_m = p_m^* f_m$

For the simple functional $K(f_m) = p_m^* f_m$ we can make use of the evaluation of the previous section and set $\lambda = 0$. As this time we want to solve an eigenvalue problem, the subsequent generalized Dirac equations read

$$\begin{cases} i \cdot \sqrt{g^{00}} \cdot \mu + g^{00} \frac{\partial}{\partial x^0} \\ i \cdot \sqrt{g^{00}} \cdot \mu - g^{00} \frac{\partial}{\partial x^0} \end{cases}_{\Omega} w\left(x^0\right) = \pm I \cdot E \cdot w\left(x^0\right) \qquad (82)$$

where simply out of curiosity we have assumed E to be a "virtual" parameter (compare what has been said with Eq. (13)). The eight possible solutions are obtained by direct integration:

$$\overset{\text{total}}{\xi_0} = x_0 + F\left(w\left(x^0\right)\right) = x_0 + \begin{cases} p_0 \cdot \left(C\,[1]^{\pm}_{\pm} \cdot e^{\pm \frac{x^0\left(E_{\pm} i_{\cdot \mu}\right)}{\sqrt{g^{00}}}} \right) \\ \\ p_0 \cdot \left(C\,[2]^{\pm}_{\pm} \cdot e^{\pm \frac{i \cdot x^0\left(E_{\pm \mu}\right)}{\sqrt{g^{00}}}} \right) \end{cases}_{\Omega} \cdot \qquad (83)$$

We have reason to believe that these solutions are connected with the leptons (electron, muon, tauon and their neutrinos). The structure of the solution and its connection with the permanent deformation of space shows an interesting pattern possibly explaining the mass differences of the leptons. Together with the assumed character of time being a quantum effect of a higher dimension (cf. [14]), we also find evidence for neutrinos having a small mass. The investigation of this solution, however, is relatively lengthy. Therefore, we will discuss the above solution in more detail in a separate section (cf. "Quantization of the Kerr Metric" or [10]).

Chapter 3

The 1D Quantum Oscillator in the Metric Picture

In this work, the author will apply a new method for the quantization of arbitrary smooth spaces on the problem of the one-dimensional harmonic oscillator with metric. It will be shown that the classical potential if directly been translated into the quantum metric picture does not provide fully satisfying solutions. This especially holds for the metric needed to generate the typical x^2 potential of the classical case. It will be demonstrated, however, that metrics of cos- or Gaussian shape provide the necessary properties for the creation of perfect ground state solutions being similar to the classical solution.

Further results are as follows:

1. Excited classical states are just superposed ground states in the new picture.
2. For higher excitations in the metric picture, even completely symmetric setups of excitations can lead to asymmetric oscillation results of the quantized metric solution.
3. Even though we only consider the one-dimensional case, we find sub-quantum states similar to the magnetic quantum numbers as known from the separation approach to the three dimensional

The Theory of Everything: Quantum and Relativity is Everywhere – A Fermat Universe
Norbert Schwarzer
Copyright © 2020 Jenny Stanford Publishing Pte. Ltd.
ISBN 978-981-4774-47-5 (Hardcover), 978-1-315-09975-0 (eBook)
www.jennystanford.com

central force potential (hydrogen atom solution) in classical quantum physics.

4. In the case of a cos- or sin-based metric, we find special ground state solutions with wiggles mirroring the underlying metric when these ground states are stretching over several cos- or sin-waves (graphs clearly show shadows of such waves on their shoulders).

5. It will also be derived that quantized states are only possible in curved spaces while a flat metric would always bring about continuum solutions. This is in perfect agreement with the classical finding where it requires potentials and a certain maximum of energy below the potential to create non-continuum solutions. With potentials curving the space in the metric picture as used here, however, it is clear that there cannot be finite quantum solutions in a flat space, because this would be the equivalent to the absence of potentials.

6. Further, by comparison with the Heisenberg uncertainty principle, it is found that the minimum quantum steps allowed for a certain system are directly dependent on the wavelength of the underlying metric of that very system.

7. Applying this finding to gravity, we will see that, taking the classical Heisenberg principle, the smallest quantum in mass would be equivalent to a Schwarzschild radius of Planck length. This, however, completely contradicts the observed existence of elementary particles with Schwarzschild radii far below the Planck length.

8. It is concluded therefore, that gravitation does not fit into the set of quantum rules and limits the other forces and interactions are following. It was suggested that gravity must be based on a certain sublevel structure lying underneath the Planck scale. This also seems to be the reason for the exceptional behavior of gravity if compared with the other basic forces. Suggestions are made with respect to this substructure and typical constants and scale sizes of this sublevel are evaluated.

The Classical Harmonic Quantum Oscillator within the Metric Picture, or the Theory of Everything

As already evaluated previously (e.g. [9, 12, 15, 16]), we can derive a rather general V_Ω vector for the one dimensional case as:

$$\hat{V}_\Omega w\left(x^0\right) = \left\{ \begin{array}{l} i\cdot\sqrt{g^{00}}\cdot\mu + g^{00}\frac{\partial}{\partial x^0} \pm E_1 \pm e^{\frac{i\pi}{2}} E_1 \pm\ldots\pm e^{\frac{(2j-1)i\pi}{2\cdot j}} E_j \pm\ldots \\ i\cdot\sqrt{g^{00}}\cdot\mu - g^{00}\frac{\partial}{\partial x^0} \pm E_1 \pm e^{\frac{i\pi}{2}} E_1 \pm\ldots\pm e^{\frac{(2j-1)i\pi}{2\cdot j}} E_j \pm\ldots \end{array} \right\}_\Omega$$

$$w\left(x^0\right) = 0 \tag{84}$$

where simply out of curiosity we have assumed the E_i to be "virtual" parameters (compare with what has been said to equation (13)). The solutions are obtained by direct integration:

$$\xi_0^{\text{total}} = x_0 + F\left(w\left(x^0\right)\right) = x_0 + \left\{ p_0\cdot\left(C\left[1\right]_\pm\cdot e^{\pm\frac{x^0\left(e^{\frac{i j\pi}{P}}E_P - i\cdot\mu\right)}{\sqrt{g^{00}}}}\right)\right\}_\Omega$$

$$= x_0 + \left\{ p_0\cdot\left(e^{\pm C\left[2\right]_\pm\left(e^{\frac{i j\pi}{P}}E_P - i\cdot\mu\right)}\cdot e^{\pm\frac{x^0\left(e^{\frac{i j\pi}{P}}E_P - i\cdot\mu\right)}{\sqrt{g^{00}}}}\right)\right\}_\Omega$$

$$= x_0 + \left\{ p_0\cdot e^{\pm\frac{\left(x^0\pm\sqrt{g^{00}}\cdot C\left[2\right]_\pm\right)\left(e^{\frac{i j\pi}{P}}E_P - i\cdot\mu\right)}{\sqrt{g^{00}}}}\right\}_\Omega. \tag{85}$$

Again, we find a wave-like solution with the integration constant playing the role of one of the two wave coordinates in an assumed two dimensional space. It is been shown in [10] that in Kerr metrics this wave can be a quantum revolution around a mass center.

Now we will see how we can obtain the Klein–Gordon harmonic quantum oscillator of our slightly more complex functional $F(w(x^0))$. Constructing the square of our V_Ω vectors as given in (84) we result in the Klein–Gordon equation for the harmonic quantum oscillator, which is also been obtained by the Higgs potential given in [15] with the setting $\lambda = 0$ and μ becoming a quadratic function of x^0, namely $\mu = \mu\left(x^0\right) = \mu_2(x^0)^2$. We will also treat this potential later in this book (c.f. section "Generalization of 'The Recipe': From $\hbar$ to the Planck Tensor"). The subsequent equation now reads

$$\Box_{\text{Metric}}w + \mu^2 w - \lambda^2 w^3 \equiv \Box_{\text{Metric}}w - F^2\left[w\right] \equiv \left[\Box_{\text{Metric}} - \hat{F}^2\right]w = 0$$

$$\mu = \mu_2\left(x^0\right)^2; \quad \lambda = 0$$

$$\Rightarrow \quad \left[\Box_{\text{Metric}} - \mu_2\left(x^0\right)^2\right]w\left(x^0\right) = 0 \tag{86}$$

and resembles the equation for the harmonic quantum oscillator.

Applying our usual V_Ω-vectorial root extraction, we obtain a more general version of equation (84), which now reads

$$\hat{V}_\Omega w\left(x^0\right) = \left\{ \begin{array}{l} i \cdot \sqrt{g^{00}} \cdot \left(\mu_0 \pm \mu_1 \cdot \sqrt{x^0} \pm \mu_2 \cdot x^0 \pm \dots\right) \pm g^{00}\frac{\partial}{\partial x^0} \\ \pm E_1 \pm e^{\frac{i\pi}{2}} E_1 \pm \dots \pm e^{\frac{(2j-1)i\pi}{2j}} E_j \pm \dots \end{array} \right\}_\Omega$$
$$\times\, w\left(x^0\right) = 0. \tag{87}$$

For the simple setting $\mu_0 = \mu_1 = 0$ and $\mu_2 \sim i \cdot \sqrt{g^{00}}$ we immediately obtain the ground state solution of the harmonic quantum oscillator. As we will see later, in the metric picture, higher-order energy states are realized by superposing ground state excitations, which also automatically explains the integer numbers for the higher-order states of the form $E_{cn} \sim (1/2 + n)$, with E_{cn} denoting the classical energy states and n gives the integer state number.

It should be noted that, as performed in [12], one could also just assume that the metric "is taking on" all the coordinate dependencies, which is giving us back our simple $F(w)$ form but demands a more complex metric component G^{00}, being a function of the coordinate x^0. Then we simply assume eigenvalues different from the ones in (87) and set:

$$\hat{V}_\Omega w\left(x^0\right) = \left\{ \frac{i}{\sqrt{G^{00}}} \pm \frac{\partial}{\partial x^0} \pm \tilde{E}_1 \pm e^{\frac{i\pi}{2}} \tilde{E}_1 \pm \dots \pm e^{\frac{(2j-1)i\pi}{2j}} \tilde{E}_j \pm \dots \right\}_\Omega w\left(x^0\right) = 0$$
$$\frac{1}{\sqrt{G^{00}}} = \left(\mu_0 \pm \mu_1 \cdot \sqrt{x^0} \pm \mu_2 \cdot x^0 \pm \dots\right). \tag{88}$$

Again, we find the ground state solution for the harmonic quantum oscillator with the assumption of a metric of the form

$$\frac{1}{\sqrt{G^{00}}} \sim i \cdot x^0 \Rightarrow \hat{V}_\Omega w\left(x^0\right)$$
$$= \left\{ -\mu \cdot x^0 \pm \frac{\partial}{\partial x^0} \pm \tilde{E}_1 \pm e^{\frac{i\pi}{2}} \tilde{E}_1 \pm \dots \pm e^{\frac{(2j-1)i\pi}{2j}} \tilde{E}_j \pm \dots \right\}_\Omega$$
$$\times\, w\left(x^0\right) = 0 \tag{89}$$

where μ has to be positive definite and the plus sign before the derivative will lead to solutions not converging for infinite x^0, which only leaves us the following equation to solve:

$$\frac{1}{\sqrt{G^{00}}} \sim i \cdot x^0 \Rightarrow \hat{V}_\Omega w\left(x^0\right)$$
$$= \left\{ -\mu \cdot x^0 - \frac{\partial}{\partial x^0} \pm \tilde{E}_1 \pm e^{\frac{i\pi}{2}} \tilde{E}_1 \pm \dots \pm e^{\frac{(2j-1)i\pi}{2j}} \tilde{E}_j \pm \dots \right\}_\Omega$$
$$\times\, w\left(x^0\right) = 0. \tag{90}$$

Thus, the remaining equation for the ground state of the harmonic oscillator will be found for a metric $G^{00} \sim -(x^0)^{-2}; G_{00} \sim -(x^0)^2$ with the subsequent equation given above. For simplicity we will now set $x^0 = x$.

With the classical solutions of the harmonic oscillator known to be [21]

$$\psi_n(x) := \left(\frac{m \cdot \omega}{\pi \cdot \hbar}\right)^{\frac{1}{4}} \frac{1}{\sqrt{2^n n!}} H_n[w] * e^{\left[\frac{-w^2}{2}\right]}; \quad w = x \cdot \sqrt{\frac{m \cdot \omega}{\hbar}} \tag{91}$$

we can backward-derive the metric from equation (88) and find

$$\frac{1}{\sqrt{G^{00}}} = \frac{1}{\psi_n} \frac{\partial \psi_n}{\partial x} = \left(\frac{2 \cdot n \cdot H_{n-1}(x)}{H_n(x)} - x\right), \tag{92}$$

and the result (with $H_n(x)$ denoting the Hermite polynomial of order n and argument x)

$$\Rightarrow \hat{V}_{n\Omega} w_n(x) = \left\{ \begin{matrix} \mu \cdot \left(\frac{2 \cdot n \cdot H_{n-1}(x)}{H_n(x)} - x\right) - \frac{\partial}{\partial x} \\ \pm \tilde{E}_1 \pm e^{\frac{i\pi}{2}} \tilde{E}_1 \pm \ldots \pm e^{\frac{(2j-1)i\pi}{2 \cdot j}} \tilde{E}_j \pm \ldots \end{matrix} \right\}_\Omega w_n(x) = 0. \tag{93}$$

Important hint: One should not mix up the classical energy eigenstates of energy E_{cn} of the classical quantum mechanical solution given as $E_{cn} = \hbar\omega\left(\frac{1}{2}+n\right)$ with our virtual eigenvalues and eigensolutions resulting from non-zero $\tilde{E}_i$! These $\tilde{E}_i$, namely, have to be understood as true virtual parameters and thus, they are the equivalent for the so-called "second quantization" resulting from the perturbations caused by the background quantum field. Here, however, this background quantum field just reveals itself as the additional jitter of the space-time adding up on the metric's curvature.

We find infinite maxima for G^{00} at those positions where the oscillator classically gives the highest probability to find oscillating particle for the various states. We also find minima ($\psi_n(x_i) = 0$), where the probability of the classical solution is zero. Already in connection with the classical result, one might ask how the particle, being trapped in the harmonic oscillator, could possible "jump" over the threshold of points of zero "probability of existence" in order to be found in a neighboring region of non-zero probability if it is not allowed to ever pass the point of $\psi_n(x_i) = 0$ in between the two regions. There was never a truly satisfying answer given to that question, except, that the theory is not complete and additional quantum effects leading to enough perturbation to somehow allow for the necessary "jump-effects." Now, however, with the classical model also being connected

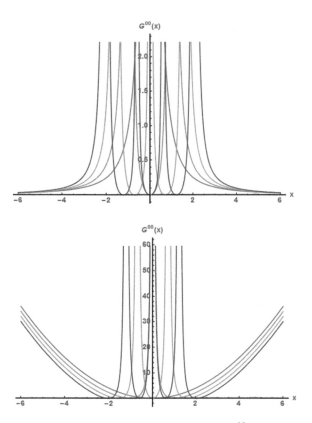

Figure 1 Solutions for the metric components G_{00} and G^{00} as given in (92) for the first four states ($n = 0, 1, 2, 3$) of the classical harmonic quantum oscillator. We used arbitrary units.

with the subsequent metric state of such an oscillator object, we are facing an even severer problem, namely the singularities occurring in connection with the classical solution.

The unfortunate aspect about our finding of the metric for the states of the classical harmonic quantum oscillator can be seen in Fig. 1. There we find singularities in the metrics G^{00} and G_{00} at various positions for our quantum object, being trapped in the oscillator. Such singularities are considered "unnatural" (as unnatural as the probability zeros) by the author and therefore we shall investigate how they could be avoided.

But before we do this, we intent to consider the corresponding metric Klein–Gordon equation (33) in order to see that an evaluation from there might help us in better understanding the singularities. In the classical case with a potential V the Klein–Gordon equation does read

$$\left[\frac{\partial_t^2}{c^2} - \Delta_{\text{Cartesian}} + \frac{M^2 c^2}{\hbar^2} + V\right]\Psi = 0. \tag{94}$$

We set this equal to our metric Klein–Gordon equation and demand the "wave function" to be identical for both equations, which gives us

$$\left[\frac{\partial_t^2}{c^2} - \Delta_{\text{Cartesian}} + \frac{M^2 c^2}{\hbar^2} + V\right]\Psi = 0 = \left[p_m^2 + \Delta_{\text{Metric}}\right]\Psi + \frac{\Psi}{4} \cdot \frac{g^{nm}}{\sqrt{g}} \partial_m \sqrt{g} \cdot g^{mi} \partial_i g_{nm}$$

$$\Rightarrow 0 = \left[\Delta_{\text{Metric}} - \frac{\partial_t^2}{c^2} + \Delta_{\text{Cartesian}}\right]\Psi + \Psi \cdot \left(\frac{g^{nm}}{4\sqrt{g}}\partial_m \sqrt{g} \cdot g^{mi} \partial_i g_{nm} - V\right)$$

$$\Rightarrow V = \frac{1}{\Psi}\left[\Delta_{\text{Metric}} - \frac{\partial_t^2}{c^2} + \Delta_{\text{Cartesian}}\right]\Psi + \left(\frac{g^{nm}}{4\sqrt{g}}\partial_m \sqrt{g} \cdot g^{mi} \partial_i g_{nm}\right). \tag{95}$$

With Ψ and the potential V not depending on time t like in the classical harmonic oscillator, we obtain (caution: we have to be aware of the sign convention for the spatial metric components in the form $\{+, -, -, -\}$, while the Cartesian Laplace operator is been given as $\Delta_{\text{Cartesian}} = \partial_x^2 + \partial_y^2 + \partial_z^2$)

$$V = \frac{1}{\Psi}\left[\Delta_{\text{Metric}} - \frac{\partial_t^2}{c^2} + \Delta_{\text{Cartesian}}\right]\Psi + \left(\frac{g^{nm}}{4\sqrt{g}}\partial_m \sqrt{g} \cdot g^{mi} \partial_i g_{nm}\right)$$

$$\underbrace{\Delta_{\text{Metric}} - \frac{\partial_t^2}{c^2} \equiv \frac{1}{\sqrt{g}}\partial_\alpha \sqrt{g} \cdot g^{\alpha\beta} \partial_\beta}_{\text{3D}-\Delta-\text{Operator}}$$

$$\Rightarrow V = \frac{1}{\Psi}\left[\underbrace{\frac{1}{\sqrt{g}}\partial_\alpha \sqrt{g} \cdot g^{\alpha\beta} \partial_\beta}_{\text{3D}-\Delta-Operator} + \Delta_{\text{Cartesian}}\right]\Psi + \left(\frac{g^{nm}}{4\sqrt{g}}\partial_m \sqrt{g} \cdot g^{mi} \partial_i g_{nm}\right)$$

$$\Rightarrow V = \frac{1}{\Psi}\left[\frac{1}{\sqrt{g}}\partial_\alpha \sqrt{g} \cdot g^{\alpha\beta} \partial_\beta + \Delta_{\text{Cartesian}}\right]\Psi + \left(\frac{g^{nm}}{4}\underbrace{\frac{1}{\sqrt{g}}\partial_\alpha \sqrt{g} \cdot g^{\alpha\beta} \partial_\beta}_{\text{3D}-\Delta-\text{Operator}} g_{nm}\right). \tag{96}$$

Now, we are going to the one-dimensional case and obtain

$$\Delta_{\text{Metric}(x)} - \frac{\partial_t^2}{c^2} \equiv \underbrace{\frac{1}{\sqrt{g}}\partial_x \sqrt{g} \cdot g^{xx} \partial_x}_{1D-\Delta-\text{Operator}}$$

$$\Rightarrow V = \frac{1}{\Psi}\left[\partial_x^2 + \frac{1}{\sqrt{g}}\partial_x \sqrt{g} \cdot g^{xx} \partial_x\right]\Psi + \left(\frac{g^{nm}}{2}\underbrace{\frac{1}{\sqrt{g}}\partial_x \sqrt{g} \cdot g^{xx} \partial_x}_{1D-\Delta-\text{Operator}} g_{nm}\right). \tag{97}$$

For a general 1D + time metric, this leads us to

$$\Rightarrow V = \frac{1}{\Psi} \left[\partial_x^2 + \frac{1}{\sqrt{g}} \partial_x \sqrt{g} \cdot g^{xx} \partial_x \right] \Psi$$

$$+ \left(\frac{\begin{pmatrix} \frac{1}{c^2} g^{x0} \\ g^{x0} g^{xx} \end{pmatrix}}{2} \underbrace{\frac{1}{\sqrt{g}} \partial_x \sqrt{g} \cdot g^{xx} \partial_x}_{1D-\Delta-\text{Operator}} \begin{pmatrix} c^2 g_{x0} \\ g_{x0} g_{xx} \end{pmatrix} \right)$$

$$\Rightarrow V = \frac{1}{\Psi} \left[\partial_x^2 + \frac{1}{\sqrt{c^2 g_{xx} - g_{x0}^2}} \partial_x \sqrt{c^2 g_{xx} - g_{x0}^2} \cdot \frac{1}{c^2 g_{xx} - g_{x0}^2} \partial_x \right] \Psi$$

$$+ \left(\frac{\begin{pmatrix} \frac{1}{c^2} g^{x0} \\ g^{x0} g^{xx} \end{pmatrix}}{2} \frac{1}{\sqrt{c^2 g_{xx} - g_{x0}^2}} \partial_x \sqrt{c^2 g_{xx} - g_{x0}^2} \cdot \frac{1}{c^2 g_{xx} - g_{x0}^2} \partial_x \begin{pmatrix} c^2 g_{x0} \\ g_{x0} g_{xx} \end{pmatrix} \right). $$

$$(98)$$

Further evaluation yields

$$\Rightarrow V = \frac{1}{\Psi} \left[\partial_x^2 + \frac{1}{\sqrt{c^2 g_{xx} - g_{x0}^2}} \partial_x \frac{1}{\sqrt{c^2 g_{xx} - g_{x0}^2}} \partial_x \right] \Psi$$

$$+ \left(\frac{\begin{pmatrix} g_{xx} & -g_{x0} \\ -g_{x0} & c^2 \end{pmatrix}}{2(c^2 g_{xx} - g_{x0}^2)} \frac{1}{\sqrt{c^2 g_{xx} - g_{x0}^2}} \partial_x \frac{1}{\sqrt{c^2 g_{xx} - g_{x0}^2}} \partial_x \begin{pmatrix} c^2 g_{x0} \\ g_{x0} g_{xx} \end{pmatrix} \right).$$

$$(99)$$

Now we set $g_{x0} = 0$:

$$\Rightarrow V = \frac{1}{\Psi} \left[\partial_x^2 + \frac{1}{c^2 \sqrt{g_{xx}}} \partial_x \frac{1}{\sqrt{g_{xx}}} \partial_x \right] \Psi$$

$$+ \left(\frac{\begin{pmatrix} \frac{1}{c^2} & 0 \\ 0 & \frac{1}{g_{xx}} \end{pmatrix}}{2} \frac{1}{c^2 \sqrt{g_{xx}}} \partial_x \frac{1}{\sqrt{g_{xx}}} \partial_x \begin{pmatrix} c^2 & 0 \\ 0 & g_{xx} \end{pmatrix} \right). \qquad (100)$$

Summing all up gives

$$\Rightarrow V = \frac{1}{\Psi}\left[\partial_x^2 + \frac{1}{c^2\sqrt{g_{xx}}}\partial_x\frac{1}{\sqrt{g_{xx}}}\partial_x\right]\Psi$$

$$+\left(\frac{\begin{pmatrix}\frac{1}{c^2} & 0 \\ 0 & \frac{1}{g_{xx}}\end{pmatrix}}{2}\frac{1}{c^2\sqrt{g_{xx}}}\partial_x\frac{1}{\sqrt{g_{xx}}}\partial_x\begin{pmatrix}c^2 & 0 \\ 0 & g_{xx}\end{pmatrix}\right). \tag{101}$$

$$\Rightarrow V = \frac{1}{\Psi}\left[\partial_x^2 + \frac{1}{c^2\sqrt{g_{xx}}}\partial_x\frac{1}{\sqrt{g_{xx}}}\partial_x\right]\Psi + \left(\frac{1}{2g_{xx}}\frac{1}{c^2\sqrt{g_{xx}}}\partial_x\frac{1}{\sqrt{g_{xx}}}\partial_x g_{xx}\right). \tag{102}$$

Now we evaluate the derivatives of the square roots of the metric components:

$$V = \frac{1}{\Psi}\left[\partial_x^2 + \frac{1}{c^2 g_{xx}}\partial_x^2 - \frac{[\partial_x g_{xx}]}{2c^2 g_{xx}^2}\partial_x\right]\Psi + \frac{1}{2g_{xx}}\frac{1}{c^2}\left(\frac{1}{g_{xx}}\partial_x^2 - \frac{[\partial_x g_{xx}]}{2g_{xx}^2}\partial_x\right)g_{xx}$$

$$= \frac{1}{\Psi}\left[\partial_x^2\Psi + \frac{\partial_x^2\Psi}{c^2 g_{xx}} - \frac{[\partial_x g_{xx}]}{2c^2 g_{xx}^2}\partial_x\Psi\right] + \frac{1}{2g_{xx}}\frac{1}{c^2}\left(\frac{\partial_x^2 g_{xx}}{g_{xx}} - \frac{[\partial_x g_{xx}]^2}{2g_{xx}^2}\right) \tag{103}$$

and get

$$\Rightarrow V = \frac{1}{\Psi}\left[\partial_x^2 + \frac{1}{c^2\sqrt{g_{xx}}}\partial_x\frac{1}{\sqrt{g_{xx}}}\partial_x\right]\Psi + \left(\frac{1}{2g_{xx}}\frac{1}{c^2\sqrt{g_{xx}}}\partial_x\frac{1}{\sqrt{g_{xx}}}\partial_x g_{xx}\right). \tag{104}$$

Reordering gives

$$V = \frac{1}{\Psi}\partial_x^2\Psi + \frac{1}{\Psi}\frac{1}{c^2}\left[\frac{1}{g_{xx}}\partial_x^2\Psi - \frac{1}{2g_{xx}^2}[\partial_x\Psi]^2\right]$$

$$+\frac{1}{2g_{xx}}\frac{1}{c^2}\left(\frac{1}{g_{xx}}\partial_x^2 g_{xx} - \frac{1}{2g_{xx}^2}[\partial_x g_{xx}]^2\right). \tag{105}$$

Here we do not intend to solve this non-linear differential equation with respect to g_{xx} but only to demonstrate that we still obtain the inconsistency with respect to diverging metric components. A simplified evaluation with settings of the form $\Psi \sim g_{xx}$ or $\frac{1}{\Psi}\frac{1}{c^2}\left[\frac{1}{g_{xx}}\partial_x^2\Psi - \frac{1}{2g_{xx}^2}[\partial_x\Psi]^2\right] + \frac{1}{2g_{xx}}\frac{1}{c^2}\left(\frac{1}{g_{xx}}\partial_x^2 g_{xx} - \frac{1}{2g_{xx}^2}[\partial_x g_{xx}]^2\right) = 0$ suffices and we get

$$V = \frac{1}{\Psi}\partial_x^2\Psi + \frac{3}{2\Psi^2}\frac{1}{c^2}\left(\partial_x^2\Psi - \frac{1}{2\Psi}[\partial_x\Psi]^2\right) \text{ or } V = \frac{1}{\Psi}\partial_x^2\Psi. \qquad (106)$$

As expected we still find singularities for the metric (approximated or not).

Thus, the metric Klein–Gordon (33) or the metric Schrödinger equation (41) will not help us in getting rid of the metric singularities. We conclude that, apparently, the classical potential picture is not complete, and therefore its direct translation into the metric picture can lead to problems.

Now we proceed with our metric Dirac approach and try to find metrics, which avoid singularities, but still contain respectively mirror at least partially harmonic potentials.

It was suggested to allow for non-virtual eigenvalues in the equation for the "intelligent zeros" (52) and/or their subsequent roots. Equation (87) would then read like this:

$$\hat{V}_\Omega w\left(x^0\right) = \left\{ \begin{array}{l} i \cdot \sqrt{g^{00}} \cdot \left(\mu_0 \pm \mu_1 \cdot \sqrt{x^0} \pm \mu_2 \cdot x^0 \pm \ldots\right) \pm g^{00}\frac{\partial}{\partial x^0} \\ +E_0 \pm E_1 \pm e^{\frac{i\pi}{2}}E_1 \pm \ldots \pm e^{\frac{(2j-1)i\pi}{2\cdot j}}E_j \pm \ldots \end{array} \right\}_\Omega$$

$$\times w\left(x^0\right) = 0 \qquad (107)$$

and one could consider E_0 being the positive and negative roots of the classical energy eigenvalues $E_0 = \pm\sqrt{E_n}$. Apart from the fact that this would contradict our conceptual approach of sucking all non-virtual "perturbations" into the metric it will also not solve our problem with the singularities for the metric of the classical potential of the harmonic oscillator. One argument could be, of course, that the quantum oscillator in only one dimension might not be consistent with all aspects of reason, because truly one-dimensional cases do not seem to exist in our universe. Even string theory usually considers its 1D objects to have a certain extension in width.

We already know that the harmonic potential of x^2 character, becoming infinite for infinite x, can be considered a weakness of the simplicity of the model. This is obvious as a monotonic slope until infinity will not be found in nature. Even the permanent descent of curvature of a classical Schwarzschild black hole currently finds its realistic end near or at least at the event horizon, while new concepts (e.g. [17]) give the quantized Schwarzschild solution a non-singular escape. Thus, more realistic potentials with converging "rims" should not result in such problems for infinite x, but what about the singularities in the middle and what about the zeros in the probability density?

Gaussian-Like Metric Approach

In order to simplify the discussion we want to investigate a potential not having such weaknesses, especially not the one with the infinite grows for infinite argument. We start with a metric of the form:

$$\frac{1}{\sqrt{G^{00}}} = \mu_0 + \mu_1 \cdot e^{-\tilde{\omega} \cdot x^2} \Rightarrow G^{00} = \left(\mu_0 + \mu_1 \cdot e^{-\tilde{\omega} \cdot x^2}\right)^{-2}. \tag{108}$$

The solutions to the subsequent differential equations according to (87) would be as follows:

$$f(x) = C \cdot e^{\pm x \cdot \mu_0 \left(e^{\frac{ij\pi}{P}} E_P \cdot \mu_0 - i\right) \pm \frac{\sqrt{\pi}\mu_1 \left(-2i \cdot \operatorname{erf}[x\sqrt{\omega}] + e^{\frac{ij\pi}{P}} E_P \cdot (4 \cdot \mu_0 \cdot \operatorname{erf}[x\sqrt{\omega}] + \sqrt{2} \cdot \mu_1 \cdot \operatorname{erf}[\sqrt{2}x\sqrt{\omega}])\right)}{4\sqrt{\omega}}}, \tag{109}$$

where erf(x) stands for the error function and C is the integrational constant. In order to demand smoothness for the symmetry point $x = 0$, we set $\frac{\partial f(x)}{\partial x}\Big|_{x=0} = 0 = e^{\frac{ij\pi}{P}} E_P (\mu_0 + \mu_1) - i$ and determine μ_0 accordingly. Figures 2 to 5 show a variety of possibilities for this solution in connection with the corresponding metric. As expected, we find that every change of state also, of course, changes the metric of our one-dimensional space. In order

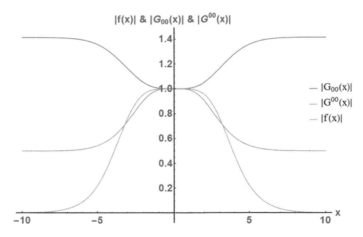

Figure 2 Solutions for the metric components G_{00} and G^{00} and the "probability density" $f(x)$ for a Gaussian-like metric approach as given in (108) and (109) for the setting $P = 2$, $E_P = 1$, $\tilde{\omega} = 0.1$, $\mu_1 = 1$. All non-converging solutions are omitted (arbitrary units). We easily recognize the ground state.

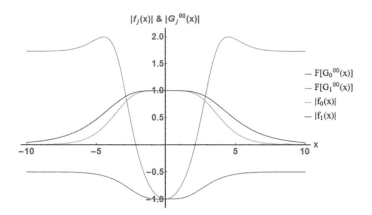

Figure 3 Solutions for the metric components G_{00} and G^{00} and the "probability density" $f(x)$ for a Gaussian-like metric approach as given in (108) and (109) for the setting $P = 3$, $E_P = 1$, $\tilde{\omega} = 0.1$, $\mu_1 = 1$. All non-converging solutions are omitted. We used arbitrary units.

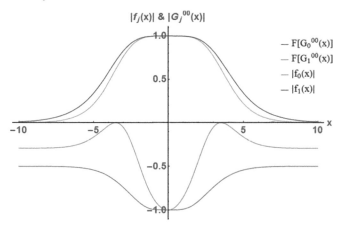

Figure 4 Solutions for the metric components G_{00} and G^{00} and the "probability density" $f(x)$ for a Gaussian-like metric approach as given in (108) and (109) for the setting $P = 4$, $E_P = 1$, $\tilde{\omega} = 0.1$, $\mu_1 = 1$. All non-converging solutions are omitted. We used arbitrary units.

to improve the illustration and to show the G^{00} right in connection with their resulting "probability densities" f, we used adaptations of the form $F[G^{00}(x)] = c_1^* G^{00}(x) + c_2$ in the figures, with suitable constants c_1 and c_2. Here we used the expression "probability densities" and put this in "...,"

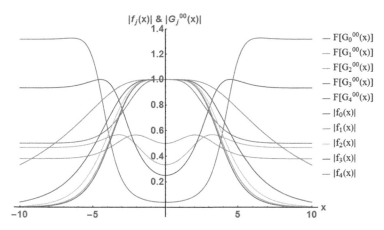

Figure 5 Solutions for the metric components G_{00} and G^{00} and the "probability density" $f(x)$ for a Gaussian-like metric approach as given in (108) and (109) for the setting $P = 9$, $E_{\mathbf{P}} = 1$, $\tilde{\omega} = 0.1$, $\mu_1 = 1$. All non-converging solutions are omitted. We used arbitrary units.

because in our picture it is less a probability than a distribution of oscillation what we have evaluated by the means of our function $f(x)$. However, we shall leave the discussion about what truly could f mean for later. We should elaborate that $\omega = \sqrt{\tilde{\omega}}$ is a true angular frequency in the case of x being the time coordinate, but it could easily also stay a frequency in the case of x denoting a spatial dimension by the simple product ω/c^*x with c denoting the speed of light in vacuum. Here, however, we will omit the factor one over c.

Situations with multiple probability maxima can easily be constructed by the means of the introduction of several centers of excitation. As various states of excitation can be combined, asymmetric solutions also are possible. This is been shown in the second set of graphs in Fig. 6. While if choosing the ground state for all centers of excitation, always a symmetric oscillation will be the result (as long as the centers of excitation are set symmetrically, of course), things are different with the excited states. There, we find strong possible asymmetry in the case of higher states of excitation (higher levels of virtual parameters) in combination with higher numbers j corresponding to virtuality sub-states, belonging to one P.

Thus, asymmetry can occur even though the original "setup" or placement of centers of excitement was completely symmetric. This becomes the

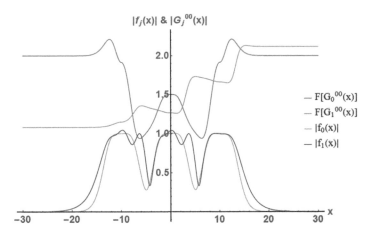

Figure 6 Solutions for the metric components G_{00} and G^{00} and the "probability density" $f(x)$ for a Gaussian-like metric approach as given in (108) and (109) with three centers of excitation at $x = -10$, 0 and 10 for the setting $P = 3$, $E_P = 1$, $\tilde{\omega} = 0.1$, $\mu_1 = 1$. All non-converging solutions are omitted. We used arbitrary units.

more pronounced the higher the levels of excitation respectively levels of "virtuality" (c.f. Fig. 7). It should be hinted, that this j-numbering seems to be similar to the various magnetic states in dependence to corresponding states of momentum. Here, however, we found it in connection with a metric 1D oscillator.

In comparison with the classical solution to the harmonic quantum oscillator, we find no singularities in the metrics. Still we are able to produce multiple maxima situations for the probability function and we obtain asymmetric situations depending on the j-number combinations for those various centers of excitation.

Cos-Like Metric Approach

Things are also interesting with yet another metric approach consisting of sin or cos functions. We now apply a basic metric of the form

$$\frac{1}{\sqrt{G^{00}}} = \mu_0 + \mu_1 \cdot \cos\left[\omega \cdot x\right] \Rightarrow G^{00} = \left(\mu_0 + \mu_1 \cdot \cos\left[\omega \cdot x\right]\right)^{-2}. \quad (110)$$

The solutions to the subsequent differential equations according to (87) would be as follows:

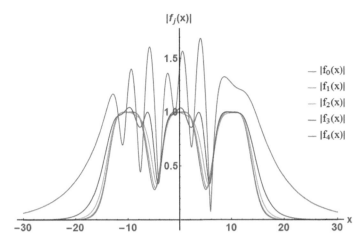

Figure 7 Solutions for the metric components G_{00} and G^{00} and the "probability density" $f(x)$ for a Gaussian-like metric approach as given in (108) and (109) with three centers of excitation at $x = -10, 0$ and 10 for the setting $P = 9$, $E_P = 1$, $\tilde{\omega} = 0.1$, $\mu_1 = 1$. All non-converging solutions are omitted. We used arbitrary units.

$$f(x) = e^{\pm \frac{-4\mathbf{i}(x\cdot\omega\cdot\mu_0+\mu_1\sin[x\cdot\omega])+e^{\frac{\mathbf{i}j\pi}{P}}E_P\left(2x\cdot\omega\left(2\mu_0^2+\mu_1^2\right)+8\cdot\mu_0\mu_1\sin[x\cdot\omega]+\mu_1^2\sin[2x\cdot\omega]\right)}{4\omega}} \cdot C. \qquad (111)$$

Applying the same boundary condition as used with the Gaussian approach, we result in similar solutions as before if setting the wavelength of our cos-approach big enough (which means ω to become small enough). Figure 8 shows the resulting function f and metric distortion for such a case.

Again, we recognize the ground state and the similarity with the classical ground state geometry (classical harmonic quantum oscillator). If, however, the wavelength is becoming smaller, the probability density might stretch (depending on the other parameters) over several metric maxima taking on strange "wiggles" along its shoulders as it can be observed in Fig. 9.

Question of Quantizing the Solution

In the classical cases, the quantization is automatically coming from boundary conditions. So, for example, the general solution for the time-independent Schrödinger equation of the harmonic oscillator would be a parabolic cylinder function $D_n(z)$ with arbitrary n. Only the fact that we

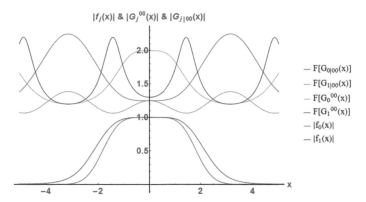

$$|f_j(x)| \ \& \ |G_j^{00}(x)| \ \& \ |G_{j|00}(x)|$$

- $F[G_{0|00}(x)]$
- $F[G_{1|00}(x)]$
- $F[G_0^{00}(x)]$
- $F[G_1^{00}(x)]$
- $|f_0(x)|$
- $|f_1(x)|$

Figure 8 Solutions for the metric components G_{00} and G^{00} and the "probability density" $f(x)$ for a cos-like metric approach as given in (110) and (111) with only one center of excitation at $x = 0$ for the setting $P = 3$, $E_P = 1$, $\omega = 1$, $\mu_1 = 1$. All non-converging solutions are omitted. We used arbitrary units.

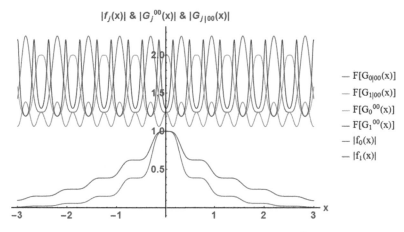

$$|f_j(x)| \ \& \ |G_j^{00}(x)| \ \& \ |G_{j|00}(x)|$$

- $F[G_{0|00}(x)]$
- $F[G_{1|00}(x)]$
- $F[G_0^{00}(x)]$
- $F[G_1^{00}(x)]$
- $|f_0(x)|$
- $|f_1(x)|$

Figure 9 Solutions for the metric components G_{00} and G^{00} and the "probability density" $f(x)$ for a cos-like metric approach as given in (110) and (111) with only one center of excitation at $x = 0$ for the setting $P = 3$, $E_P = 1$, $\omega = 10$, $\mu_1 = 1$. All non-converging solutions are omitted. We used arbitrary units.

demand the solution to behave converging with the argument z going to plus or minus infinity restricts it to integer n and thus, gives us the classical solution with the Hermite polynomials times a Gaussian factor. In our solutions within the sections above, we already have found finite, which is to say non-singularity solutions without resorting to quantization methods. It might well be, that the author so far has simply overseen a boundary condition also within the metric picture, allowing for the usual quantization, but from the current position it appears adequate to assume that in connection with certain limiting scales such boundary conditions might arise. Taking the quantization from the classical harmonic oscillator which is giving the energy eigenstates as $E_{cn} = \hbar\omega \left(\frac{1}{2} + n\right)$ ($n \ldots$ integer), we might like to find the connection via the angular frequency ω. As our approach is completely geometrical, we require an equivalent quantization condition just containing geometrical parameters. As it was already suggested by Stamler and Co. [2], we suspect the connection via the Schwarzschild radius $r_S = \frac{2 \cdot M \cdot G}{c^2}$ ($G \ldots$ gravitational constant; $M \ldots$ mass of the object) and the change of space-time volume caused by the physical process or state or massive particle (which is also just a state of oscillation) in question. We write the Heisenberg uncertainty principle in the following form:

$$\Delta x \cdot \Delta \left(M \cdot \dot{x}\right) = \Delta \omega \cdot \Delta \left(M \cdot x^2\right) = \frac{\Delta \omega}{2\pi} \cdot \Delta \left(\frac{r_s c^2}{2 \cdot G} \cdot x^2\right)$$

$$= \frac{c^2}{4\pi \cdot G} \Delta \omega \cdot \Delta V \geq \frac{\hbar}{2} \Rightarrow \frac{c^2}{G} \Delta \omega \cdot \Delta V \geq h. \quad (112)$$

Keeping the angular frequency constant and treating the uncertainty principle like an equality $\frac{c^2}{G} \Delta \omega \cdot \Delta V = \frac{c^2}{G} \omega \cdot \Delta V = h$, our quantum energies E_n for the classical harmonic oscillator becomes

$$E_{cn} = \frac{\hbar \cdot h \cdot G}{\Delta V \cdot c^2} \cdot \left(\frac{1}{2} + n\right) = \frac{h^2 \cdot G}{2\pi \cdot \Delta V \cdot c^2} \cdot \left(\frac{1}{2} + n\right). \quad (113)$$

This leads us to the suspicion that the change of volume of a certain state, process or excitation determines the energy connected with the physical situation. Further driving of the idea forward, we could even substitute the angular frequency by the corresponding wavelength λ and obtain the uncertainty principle to become

$$\frac{c^2}{G} \Delta \frac{2\pi}{\lambda} \cdot \Delta V \geq h \quad \Rightarrow \quad \frac{c^3}{G} \Delta \frac{1}{\lambda} \cdot \Delta V = \frac{c^3}{G} \cdot \frac{\Delta V}{\Delta \lambda} \geq \hbar. \quad (114)$$

In connection with our geometric solution, one could interpret (113) as the superposition of various excitations on a dimension of a certain metric. The number of excitations n and their individual volume changes determines the energy level E_n.

With respect to (114) in connection with our examples (110) and (108) our interpretation is as follows. In a flat metric the wavelength is infinite and arbitrarily small volume and subsequently mass changes are possible (continuum solutions). Within a curved metric, however, the characteristic wavelength of the curvature λ determines the minimum changes of volume to be of the kind $\Delta V \geq \hbar \cdot \lambda \cdot \frac{G}{c^3}$. This finding is not much different from the classical one where continuum solutions occur in spaces free of potentials (or energy levels above that potential) while quantized solutions are found in the presence of potentials. As in our picture these potentials are being sucked into the metric, a potential-free metric, which is flat, must, of course, result in continuum solutions, while a curved space does provide boundaries leading to quantized solutions.

With the wavelength only being another (one-dimensional) measure for the curvature, we should state that the curvature radius ρ determines the quantization conditions for the space determined by a certain metric with curvature. Using the curvature scalar S being $S = 2(\rho_1 \cdot \rho_2)^{-1}$ with the two principle curvature radii ρ_1 and ρ_2 for a surface curved into the R^3, we can substitute the wavelength. Following the example of Planck we simply introduce another "help" constant and name it $\hbar\!\!\!/$ and now substitute the wavelength by the curvature parameter S multiplied by the new "help" constant. The quantum uncertainty is now been given by

$$\frac{c^3}{G} \Delta \frac{1}{\lambda} \cdot \Delta V \quad \Rightarrow \quad \frac{c^3}{G} \Delta \sqrt{S} \cdot \hbar\!\!\!/ \cdot \Delta V = \frac{c^3}{G} \cdot \Delta \sqrt{S} \cdot \Delta V \geq \frac{\hbar}{\hbar\!\!\!/}. \tag{115}$$

In the one-dimensional case, the volume of a line should simply be the length of the line determined by our solution $f(x)$ and so we have to evaluate the integral

$$\Delta V_{1D} = \int\limits_{-\infty}^{+\infty} \sqrt{dx^2 + df^2} - \int\limits_{-\infty}^{+\infty} dx = \int\limits_{-\infty}^{+\infty} \left(\sqrt{1 + (f')^2} - 1 \right) dx. \tag{116}$$

Unfortunately, the author was not able to find analytical solutions for the integral in (116) for the Gauss- and cos-metric approaches used here and thus, only numerical results can be used for further consideration. This, however, will be done elsewhere in connection with a somewhat more practically relevant 3D metric [17].

Together with the results obtained above, we now have a way to estimate the minimum quantum portions in dependence on the spatial curvature. It will be shown elsewhere [17] that the minimum radius of curvature for the governing metric of a massive object is not getting smaller than the Schwarzschild radius of that object. This even holds true if we are talking about black holes. Thus, the minimum quantum portions, which are possible in a certain metric, are given by the maximum curvature regions or minimum "wave length areas" residing within that metric. From the Heisenberg uncertainty principle, using our considerations above, we can derive the following connection

$$\Delta V \geq \hbar \cdot \lambda_{\min} \cdot \frac{G}{c^3} \;\Rightarrow\; \sqrt[3]{\Delta V} = \Delta R_S \geq \frac{\sqrt[3]{\hbar \cdot \lambda_{\min} \cdot G}}{c} \approx \frac{\sqrt[3]{\hbar \cdot r_S \cdot G}}{c} \tag{117}$$

which is giving us an idea about Heisenberg-compatible portions of Schwarzschild radii in dependence on the metric wave length. Setting $\lambda = r_S$, which is to say, assuming the smallest possible metrical curvatures of the size of the Schwarzschild radius as set in (117), we are able to investigate the quantum options for differently sized objects (objects with different mass densities).

The result for this simple evaluation is shown in Fig. 10. What we find is rather surprising. For bigger objects (right-hand side resp. starting from about Log[λ] $= -20$) we find, as expected, always miniscule quantum portions in comparison with the total mass allowing small changes of mass in comparison to the object's original mass. Alternatively, if we want to formulate it purely geometrically, for objects with Schwarzschild radii r_S bigger than a certain size L_{crit} (s. b.) we find minimum quantum portions

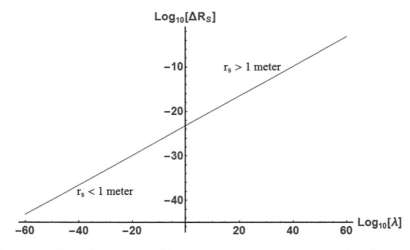

Figure 10 Heisenberg compatible minimum quantum portions for changes of Schwarzschild radius ΔR_S in dependence on the object's metric minimum wavelength $\lambda = \lambda_{\min}$.

for changes of that Schwarzschild radius ΔR_S smaller than these very radii r_S. This is a finding, which is in perfect agreement with our everyday experience and our usual experimental observations. However, for objects with Schwarzschild radii r_S **smaller** than a certain size L_{crit} (s. b.) we find minimum quantum portions for changes of that Schwarzschild radius ΔR_S **bigger** than these very radii r_S. This does not only seem to be an extremely strange, if not to say ridiculous result, it would also stand in total contrast to standard physics. Before we want to discuss this, however, we want to evaluate the limiting size L_{crit} at which the possible (Heisenberg-compatible) change of the Schwarzschild radius ΔR_S exceeds the original object's Schwarzschild radius r_S. For this, we have to solve the following equation with respect to L_{crit}:

$$L_{crit} = \frac{\sqrt[3]{\hbar \cdot L_{crit} \cdot G}}{c} \quad \Rightarrow \quad L_{crit} = 1.616229 \cdot 10^{-35} m = L_{Planck}, \qquad (118)$$

which, interestingly, is giving us the Planck length as result. Any object with a Schwarzschild radius bigger than the Planck length can easily, which is to say in a Heisenberg-compatible manner, change its mass with portions smaller than its own mass.

But what about objects with Schwarzschild radii smaller than the Planck length? As the Schwarzschild radius equal to the Planck length coincides with the Planck mass, this is a relatively huge number, especially if we are talking about known elementary particles. So, the Schwarzschild radius of the electron is only about $1.35*10^{-57}$, which is 22 orders below the Planck length. Thus, something must be missing, because, as it seems, the Heisenberg uncertainty principle does not seem to be applicable to gravity.

The Level Underneath

There are two possible explanations for the apparent strange finding in the section above:

- either what was evaluated here is incorrect at some point or
- gravitation is not "really of this world"

While we are leaving the prove of the first possible explanation to others, we here want to investigate the second possibility, namely that gravity is not part of the usual quantum world respectively that it does not follow the same rules as the other known forces and interactions. As already suggested in previous papers [1, 2, 6, 14], one could understand the coming about of the quantum effects as the result of a grain-like structure of space. The characteristic radii of such grains are of Planck size and, as already suggested, they are mostly universes of Friedmann Robertson Walker FRW style [6]. In [1] they were named "Friedmanns" and we want to keep that expression. This is the substructure of our space and, as shown before, it explains most of the quantum effects we are observing with the exception of things having to do with gravity. With respect to gravity, however, things are slightly different, because gravity is based on the volumetric deformation of the Friedmanns and this immediately brings up the question about the structure of the little Friedmanns themselves. Again, as suggested before, we think that those Friedmanns building our space are just themselves made out of even smaller Friedmann universes. We might name them "Friedmanns of the level underneath" or "Friedmanns of the second level." A successive formation process on various ever-bigger scales, as suggested in [22, 23], might be the reason for the existence of such a Matryoshka-like space. At any case, by placing the basis for gravity on the interaction of these "Friedmanns of the level underneath" and by taking the Heisenberg principle for quantum effects of our level as being based on the first level Friedmanns of our space, we immediately find a reason for the apparent sub-quantum (taking the Planck level) behavior of gravitation.

While the size of Friedmanns of the first level (ours) must be around the Planck length, it would be interesting to estimate also the size of the sub-Friedmanns. Taking the results of the LHC about the Higgs particle measurements the author already has evaluated that in the case of a two-sphere structure of all Friedmanns, the size of the second-level Friedmanns must lie in the range of about $L_P^2/m = 2.61*10^{-70}$m [23] (m ... meter). Applying (118) on this sub-Planck level we can evaluate the combined values for some sublevel constants:

$$L_{crit2} = \frac{\sqrt[3]{\hbar_{level2} \cdot L_{crit2} \cdot G_{level2}}}{c_{level2}}$$

with
$$L_{crit2} = \left(1.616229 \cdot 10^{-35}\right)^2 m = \frac{L_{Planck}^2}{m}$$

$$\Rightarrow \quad \frac{\sqrt[3]{\hbar_{level2} \cdot G_{level2}}}{c_{level2}} = 4.0863 \cdot 10^{-47} m^{2/3} \Rightarrow \frac{\hbar_{level2} \cdot G_{level2}}{c_{level2}^3}$$

$$= 6.823 \cdot 10^{-140} m^2 = \frac{L_{Planck}^4}{m^2}. \tag{119}$$

Taking now the gravitational constant of level two to be the same as of the level above, we must either have a much bigger speed of light at the sublevel two and/or a much smaller Planck constant. It will be shown elsewhere that we probably have both [17] (see also section "The Quantized Schwarzschild Metric").

Conclusions to the "Einstein Oscillator"

In this work, the author has applied the method of quantization the line element of an arbitrary space with smooth metric on the problem of the one-dimensional harmonic oscillator. It was shown that the classical potential if directly been translated into the quantum metric picture does not provide fully satisfying solutions. This especially holds for the metric needed to generate the typical x^2 potential of the classical case. It was demonstrated, however, that metrics of cos- or Gaussian shape provide the necessary properties for the creation of perfect ground state solutions being similar to the classical solution.

Further results are as follows:

1. Excited classical states are just superposed ground states in the new picture.
2. For higher excitations in the metric picture, even completely symmetric setups of excitations can lead to asymmetric oscillation results of the quantized metric solution.
3. Even though we only considered the one-dimensional case, we found sub-quantum states similar to the magnetic quantum numbers as known from the separation approach to the three dimensional central force potential (hydrogen atom solution) in classical quantum physics.
4. In the case of a cos- or sin-based metric, we found special ground state solutions with wiggles mirroring the underlying metric when these ground states are stretching over several cos- or sin-waves (graphs clearly have shown shadows of such waves on their shoulders).
5. It was also found that quantized states are only possible in curved spaces while a flat metric would always bring about continuum solutions. This is in perfect agreement with the classical finding where it requires potentials and a certain maximum of energy below the potential to create non-continuum solutions. With potentials curving

the space in the metric picture as used here, however, it is clear that there cannot be finite quantum solutions in a flat space, because this would be the equivalent to the absence of potentials.

6. Further, by comparison with the Heisenberg uncertainty principle, it was found that the minimum quantum steps allowed for a certain system are directly dependent on the wavelength of the underlying metric of that very system.

7. Applying this finding to gravity, we found that, taking the classical Heisenberg principle, the smallest quantum in mass would be equivalent to a Schwarzschild radii of the Planck length. This, however, completely contradicts the observed existence of elementary particles with Schwarzschild radii far below the Planck length.

8. It was concluded therefore, that gravitation does not fit into the set of quantum rules and limits, which the other forces and interactions are following within our spatial scale level. It was suggested that gravity must be based on a certain sublevel structure lying underneath the Planck scale. This also seems to be the reason for the exceptional behavior of gravity if compared with the other basic forces. Suggestions have been made with respect to this substructure and typical constants and scale sizes of this sublevel were evaluated.

Chapter 4

The Quantized Schwarzschild Metric

Already in [7] it was shown how to apply the quantization method described in the previous sections on the Schwarzschild metric, which can be given as

$$g^{\alpha\beta}_{\text{Schwarzschild}} = \begin{pmatrix} \frac{1}{c^2}\left(1 - \frac{2\cdot G\cdot m}{c^2\cdot r}\right)^{-1} & 0 & 0 & 0 \\ 0 & \frac{2\cdot G\cdot m}{c^2\cdot r} - 1 & 0 & 0 \\ 0 & 0 & -\frac{1}{r^2} & 0 \\ 0 & 0 & 0 & -\frac{1}{r^2\cdot\sin^2\vartheta} \end{pmatrix}. \tag{120}$$

Therefore, here we will only give the differential metric equations and their solutions. As shown in connection with (13), we could introduce virtual parameters ε without changing the result for the scalar product of our V_Ω vectors. Subsequently, we obtain the following differential equations of first order with respect of the g^{00} component:

$$0 = \begin{pmatrix} -\mu\sqrt{g^{00}}\,f_m + i\cdot g^{00}\partial_0 f_0 - i\cdot\varepsilon^m f_m \\ -\mu\sqrt{g^{00}}\,f_m + i\cdot g^{00}\partial_0 f_0 + i\cdot\varepsilon^m f_m \\ -\mu\sqrt{g^{00}}\,f_m - i\cdot g^{00}\partial_0 f_0 - \varepsilon^m f_m \\ -\mu\sqrt{g^{00}}\,f_m - i\cdot g^{00}\partial_0 f_0 + \varepsilon^m f_m \end{pmatrix}; \quad m = 0. \tag{121}$$

A similar set of equations is been obtained regarding the g^{11} component:

$$0 = \begin{pmatrix} -\lambda\sqrt{g^{11}}\,f_m + i\cdot g^{11}\partial_1 f_1 - i\cdot\varepsilon^m f_m \\ -\lambda\sqrt{g^{11}}\,f_m + i\cdot g^{11}\partial_1 f_1 + i\cdot\varepsilon^m f_m \\ -\lambda\sqrt{g^{11}}\,f_m - i\cdot g^{11}\partial_1 f_1 - \varepsilon^m f_m \\ -\lambda\sqrt{g^{11}}\,f_m - i\cdot g^{11}\partial_1 f_1 + \varepsilon^m f_m \end{pmatrix}; \quad m = 1. \tag{122}$$

The Theory of Everything: Quantum and Relativity is Everywhere – A Fermat Universe
Norbert Schwarzer
Copyright © 2020 Jenny Stanford Publishing Pte. Ltd.
ISBN 978-981-4774-47-5 (Hardcover), 978-1-315-09975-0 (eBook)
www.jennystanford.com

Here are the two sets of solutions (with $\varepsilon = 0$):

$$\begin{pmatrix} f_{(0)}(t,r) \\ f_{(1)}(r) \end{pmatrix} = \left\{ \begin{pmatrix} C_{01} \cdot e^{\frac{-i \cdot t \cdot c \cdot \mu}{\sqrt{\frac{r}{(r-r_s)}}}} \\ C_{11} \cdot h_1(r) \end{pmatrix}, \begin{pmatrix} C_{02} \cdot e^{\frac{-i \cdot t \cdot c \cdot \mu}{\sqrt{\frac{r}{(r-r_s)}}}} \\ C_{12} \cdot h_2(r) \end{pmatrix}, \right.$$

$$\left. \times \begin{pmatrix} C_{03} \cdot e^{\frac{i \cdot t \cdot c \cdot \mu}{\sqrt{\frac{r}{(r-r_s)}}}} \\ C_{13} \cdot h_1(r) \end{pmatrix}, \begin{pmatrix} C_{04} \cdot e^{\frac{i \cdot t \cdot c \cdot \mu}{\sqrt{\frac{r}{(r-r_s)}}}} \\ C_{14} \cdot h_2(r) \end{pmatrix} \right\}$$

$$h_1(r) = e^{i \cdot r \cdot \lambda \sqrt{\frac{r_s}{r}-1} - \frac{\lambda}{2} i \cdot r_s \operatorname{ArcCot}\left[\frac{2r\sqrt{\frac{r_s}{r}-1}}{2r-r_s}\right]};$$

$$h_2(r) = e^{-i \cdot r \cdot \lambda \sqrt{\frac{r_s}{r}-1} + \frac{\lambda}{2} i \cdot r_s \operatorname{ArcCot}\left[\frac{2r\sqrt{\frac{r_s}{r}-1}}{2r-r_s}\right]}$$

$$r_s = \frac{2GM}{c^2}. \tag{123}$$

The solutions with non-zero ε can easily be obtained by following the examples given in [9]. From there we can also extract an interesting generalization with respect to virtual parameters of the form

$$\begin{pmatrix} -\mu\sqrt{g^{00}} + i \cdot g^{00}\partial_0 \pm E_1 \pm e^{\frac{i\pi}{2}} E_1 \pm \ldots \pm e^{\frac{(2j-1)i\pi}{2 \cdot j}} E_j \pm \ldots \\ -\mu\sqrt{g^{00}} + i \cdot g^{00}\partial_0 \pm E_1 \pm e^{\frac{i\pi}{2}} E_1 \pm \ldots \pm e^{\frac{(2j-1)i\pi}{2 \cdot j}} E_j \pm \ldots \end{pmatrix} f_0 = 0 \tag{124}$$

and similarly for the contravariant metric g^{11} component

$$\begin{pmatrix} -\lambda\sqrt{g^{11}} + i \cdot g^{11}\partial_1 \pm \tilde{E}_1 \pm e^{\frac{i\pi}{2}} \tilde{E}_1 \pm \ldots \pm e^{\frac{(2j-1)i\pi}{2 \cdot j}} \tilde{E}_j \pm \ldots \\ -\lambda\sqrt{g^{11}} + i \cdot g^{11}\partial_1 \pm \tilde{E}_1 \pm e^{\frac{i\pi}{2}} \tilde{E}_1 \pm \ldots \pm e^{\frac{(2j-1)i\pi}{2 \cdot j}} \tilde{E}_j \pm \ldots \end{pmatrix} f_1 = 0. \tag{125}$$

The solutions are

$$\begin{pmatrix} f_{(0)}(t,r) \\ f_{(1)}(r) \end{pmatrix} = \left\{ \begin{pmatrix} C_{01} \cdot g_1(t) \\ C_{11} \cdot h_1(r) \end{pmatrix}, \begin{pmatrix} C_{02} \cdot g_1(t) \\ C_{12} \cdot h_2(r) \end{pmatrix}, \right.$$

$$\left. \begin{pmatrix} C_{03} \cdot g_2(t) \\ C_{13} \cdot h_1(r) \end{pmatrix}, \begin{pmatrix} C_{04} \cdot g_2(t) \\ C_{14} \cdot h_2(r) \end{pmatrix} \right\}$$

$$g_1(t) = e^{-\frac{i \cdot c^2 \left(e^{\frac{i \cdot j \cdot \pi}{p}} E_p + \mu \sqrt{\frac{r}{c^2(r-r_s)}} \right) \cdot (r-r_s) \cdot t}{r}};$$

$$g_2(t) = e^{\frac{i \cdot c^2 \left(e^{\frac{i \cdot j \cdot \pi}{p}} E_p + \mu \sqrt{\frac{r}{c^2(r-r_s)}} \right) \cdot (r-r_s) \cdot t}{r}};$$

$$h_1(r) = e^{i \cdot r \cdot \lambda \sqrt{\frac{r_s}{r}-1} - \frac{\lambda}{2} i \cdot r_s \operatorname{ArcCot}\left[\frac{2r\sqrt{\frac{r_s}{r}-1}}{2r-r_s}\right] - i \cdot e^{\frac{i \cdot j \cdot \pi}{p}} E_p \cdot r} \cdot (r - r_s)^{-i \cdot e^{\frac{i \cdot j \cdot \pi}{p}} E_p \cdot r_s};$$

$$h_2(r) = e^{-i \cdot r \cdot \lambda \sqrt{\frac{r_s}{r} - 1} + \frac{1}{2} i \cdot r_s \text{ArcCot}\left[\frac{2r\sqrt{\frac{r_s}{r}-1}}{2r - r_s}\right] - i \cdot e^{\frac{i \cdot j \cdot \pi}{p}} E_P \cdot r} \cdot (r - r_s)^{-i \cdot e^{\frac{i \cdot j \cdot \pi}{p}} E_P \cdot r_s};$$

$$r_s = \frac{2GM}{c^2}. \tag{126}$$

The Quantization of Time in the Vicinity of a Schwarzschild Object

Excluding the solutions which contain singularities (e.g. for $t > 0$ this would obviously be $g_2(t)$), we want to investigate the r-behavior of $f_0(t, r)$ at various t. In order to get a feeling for the behavior in the infinity, we evaluate the limit of $f_0(t, r)$ with r to ∞:

$$f_{(0)}(t, r) = \{C_{01} \cdot g_1(t), C_{02} \cdot g_1(t), C_{03} \cdot g_2(t), C_{04} \cdot g_2(t)\}$$

$$\lim_{r \to \infty} g_1 = \lim_{r \to \infty} e^{-\frac{i \cdot c^2 \left(e^{\frac{i \cdot j \cdot \pi}{p}} E_P + \mu \sqrt{\frac{r}{c^2(r - r_s)}}\right) \cdot (r - r_s) \cdot t}{r}}$$

$$= e^{-i \cdot c^2 t \left(e^{\frac{i \cdot j \cdot \pi}{p}} E_P + \frac{\mu}{c}\right)} \xrightarrow{E_P = 0} e^{-i \cdot c \cdot t \cdot \mu}$$

$$\lim_{r \to \infty} g_2 = \lim_{r \to \infty} e^{\frac{i \cdot c^2 \left(e^{\frac{i \cdot j \cdot \pi}{p}} E_P + \mu \sqrt{\frac{r}{c^2(r - r_s)}}\right) \cdot (r - r_s) \cdot t}{r}}$$

$$= e^{i \cdot c^2 t \left(e^{\frac{i \cdot j \cdot \pi}{p}} E_P + \frac{\mu}{c}\right)} \xrightarrow{E_P = 0} e^{i \cdot c \cdot t \cdot \mu}. \tag{127}$$

As for $E_P = 0$ we should expect to obtain the classical Dirac solutions [5] and we are able to determine the parameter μ as

$$\mu = \frac{c \cdot M}{\hbar}. \tag{128}$$

Now we set the boundary condition that n times $1/2$ number of oscillations must be finished at $r = \infty$ for all t ($n \ldots$ integer or G). This gives us the following restriction:

$$e^{\pm i \cdot c^2 t \left(e^{\frac{i \cdot j \cdot \pi}{p}} E_P + \frac{\mu}{c}\right)} = e^{\pm i \cdot n \cdot \pi} \xrightarrow{with\ n \in G} t_n = \frac{n \cdot \pi}{c^2 \left(e^{\frac{i \cdot j \cdot \pi}{p}} E_P + \frac{\mu}{c}\right)}$$

$$\Rightarrow t_n = \frac{n \cdot \pi}{c^2 \left(e^{\frac{i \cdot j \cdot \pi}{p}} E_P + \frac{M}{\hbar}\right)} = \frac{n \cdot \pi}{c^2 \left(e^{\frac{i \cdot j \cdot \pi}{p}} E_P + \frac{c^2 \cdot r_s}{2 \cdot G \cdot \hbar}\right)} \tag{129}$$

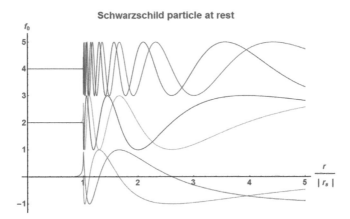

Figure 11   Real (blue, green, violet) and imaginary parts (yellow, red, brown) of the radial quantum field component f_0 for various $t = 1, 2, 5$. The Schwarzschild radius r_s was kept constant at 1 ($\mu = c = 1$ and $E_P = 0$). In order to illustratively separate the fields for the various time-sets, we added a constant of 2 in the case $t = 2$ and 4 at $t = 5$.

which automatically quantizes time in the vicinity of a Schwarzschild object. Figure 11 shows a selection of states at different times in the vicinity of the origin and the Schwarzschild radius.

The Quantization of Mass for a Schwarzschild Object

Interestingly, we also find oscillations inside the Schwarzschild radius with respect to our $f_1(r)$ solution, which is to say for $r < r_s$. Because the author was not able to find limits of $r = r_s$ in the case of $E_P \neq 0$, we here simplify our solutions to $E_P = 0$. Again demanding, that at $r = r_s$ n times a half number of oscillations must be finished, leads us to the boundary condition:

$$\lim_{r \to r_s} h_1|_{E_P=0} = \lim_{r \to r_s} e^{i \cdot r \cdot \lambda \sqrt{\frac{r_s}{r}-1} - \frac{\lambda}{2} i \cdot r_s \text{ArcCot}\left[\frac{2r\sqrt{\frac{r_s}{r}-1}}{2r-r_s}\right]}$$

$$= e^{-\frac{i}{4}\pi \cdot r_s \lambda} \xrightarrow{\lambda = \mu = \frac{cM}{\hbar}} e^{-\frac{i}{4}\pi \cdot r_s \frac{r_s c^3}{2G\hbar}} \equiv e^{-i\pi \cdot n}$$

$$\lim_{r \to r_s} h_2|_{E_P=0} = \lim_{r \to r_s} e^{-i \cdot r \cdot \lambda \sqrt{\frac{r_s}{r}-1} + \frac{\lambda}{2} i \cdot r_s \text{ArcCot}\left[\frac{2r\sqrt{\frac{r_s}{r}-1}}{2r-r_s}\right]}$$

$$= e^{\frac{i}{4}\pi \cdot r_s \lambda} \xrightarrow{\lambda = \mu} e^{\frac{i}{4}\pi \cdot r_s \frac{r_s c^3}{2G\hbar}} \equiv e^{i\pi \cdot n}. \tag{130}$$

which then results in

$$r_s = 2\sqrt{\frac{2 \cdot n \cdot G \cdot \hbar}{c^3}}. \tag{131}$$

The Level Underneath (see also [16] or Section "The 1D Quantum Oscillator in the Metric Picture")

Thus, we have a quantization of the Schwarzschild radius and subsequently the mass. As found previously with simpler methods [16], the resulting minimum radii are above the Planck length. We evaluated for the first 20 n the following results:

$$\frac{r_s}{L_{\text{Planck}}} = \begin{Bmatrix} 2.828, \mathbf{4}, 4.899, 5.657, 6.325, 6.928, 7.483, \mathbf{8}, 8.485, 8.944, 9.381, \\ 9.798, 10.198, 10.583, 10.954, 11.314, 11.661, \mathbf{12}, 12.329, 12.649 \end{Bmatrix}$$

$$L_{\text{Planck}} = 1.616229 \cdot 10^{-35} m, \tag{132}$$

where we note the integer times 4 series (see bold numbers above). It is clear that this result completely contradicts the existence of the elementary particles with Schwarzschild radii far below the Planck length. Applying the previously obtained combination of new natural constants at smaller scales and the hypothesis of gravity not following the Heisenberg principle on our scale, but on a much smaller one (c.f. [16]), we should apply the constants

with $\quad L_{\text{Planck level2}} = \left(1.616229 \cdot 10^{-35}\right)^2 m = \dfrac{L_{\text{Planck}}^2}{m}$

$$\Rightarrow \frac{\sqrt[3]{\hbar_{\text{level2}} \cdot G_{\text{level2}}}}{c_{\text{level2}}} = 4.0863 \cdot 10^{-47} m^{2/3}$$

$$\Rightarrow \frac{\hbar_{\text{level2}} \cdot G_{\text{level2}}}{c_{\text{level2}}^3} = 6.823 \cdot 10^{-140} m^2 = \frac{L_{\text{Planck}}^4}{m^2}$$

$$= L_{\text{Planck level2}}^2. \tag{133}$$

This gives us the same numbers as already obtained in (132) but on a much smaller scale:

$$\frac{r_s}{L_{\text{Planck level2}}} = \begin{Bmatrix} 2.828, \mathbf{4}, 4.899, 5.657, 6.325, 6.928, 7.483, \mathbf{8}, 8.485, 8.944, 9.381, \\ 9.798, 10.198, 10.583, 10.954, 11.314, 11.661, \mathbf{12}, 12.329, 12.649 \end{Bmatrix}$$

$$L_{\text{Planck level2}} = \left(1.616229 \cdot 10^{-35}\right)^2 m. \tag{134}$$

Investigations in Connection with the Speed of Light within the Level Underneath

In the following, we are going to use the abbreviation PL2 for "Planck level2." Now we want to combine the quantum states for time and mass, respectively the Schwarzschild radius r_s in order to see what state combinations are possible. With only the quantized Schwarzschild radius r_s and the time steps available, we are trying for a combination of those two. We explicitly point out that this is nothing more than a trial and error study. The connector we want to apply shall be the speed of light. Unfortunately, we find the following relation:

$$r_{sn_r}(n_r) = 2\sqrt{\frac{2 \cdot n_r \cdot \hbar_{\text{level2}} \cdot G_{\text{level2}}}{c_{\text{level2}}^3}} = 2\frac{L_{\text{Planck}}^2}{m}\sqrt{2 \cdot n_r} = 2 \cdot L_{\text{PL2}}\sqrt{2 \cdot n_r}$$

$$t_{n_t}(n_t, r_s) = \frac{n_t \cdot \pi}{c_{\text{level2}}\left(\frac{c_{\text{level2}}^3 \cdot r_{sn_r}(n_r)}{2 \cdot \hbar_{\text{level2}} \cdot G_{\text{level2}}}\right)} = \frac{n_t \cdot \pi}{c_{\text{level2}}\left(\frac{2 \cdot L_{\text{PL2}}\sqrt{2 \cdot n_r}}{2 \cdot L_{\text{PL2}}^2}\right)} = \frac{n_t \cdot \pi}{c_{\text{level2}}\left(\frac{\sqrt{2 \cdot n_r}}{L_{\text{PL2}}}\right)}$$

$$\Rightarrow \frac{r_{sn_r}(n_r)}{c_{\text{level2}}} \equiv t_{n_t}(n_t, r_s) \Rightarrow \frac{2 \cdot L_{\text{PL2}}\sqrt{2 \cdot n_r}}{c_{\text{level2}}} \equiv \frac{n_t \cdot \pi}{c_{\text{level2}}\left(\frac{2 \cdot \sqrt{2 \cdot n_r}}{2 \cdot L_{\text{PL2}}}\right)}$$

$$\Rightarrow 4 \cdot n_r \equiv n_t \cdot \pi \tag{135}$$

which can never be fulfilled for pure integers n_r and n_t. We conclude that our boundary condition for t, which demands whole halves of oscillations over the whole space, which is to say until $r = $ infinity cannot be correct. We assume that there must be a finite limit we here name r_{end} for which the same boundary combined condition as used above should be defined. Setting—for simplicity—again all $E_P = 0$ we have to rewrite (129), which gives us

$$\Rightarrow t_n = \frac{n \cdot \pi}{c\left(\frac{c^3 \cdot r_s}{2 \cdot G \cdot \hbar}\right)}\sqrt{\frac{r_{\text{end}}}{(r_{\text{end}} - r_s)}} \tag{136}$$

and setting in the level2-constants we obtain

$$\Rightarrow t_{n_t} = \frac{n_t \cdot \pi}{c_{\text{level2}}\left(\frac{r_s}{2 \cdot L_{\text{PL2}}^2}\right)}\sqrt{\frac{r_{\text{end}}}{(r_{\text{end}} - r_s)}}, \tag{137}$$

which finally results in the boundary condition as it was evaluated before, but now it becomes

$$\Rightarrow \frac{r_{s n_r}(n_r)}{c_{\text{level2}}} \equiv t_{n_t}(n_t, r_s) \Rightarrow \frac{2 \cdot L_{\text{PL2}} \sqrt{2 \cdot n_r}}{c_{\text{level2}}}$$

$$\equiv \frac{n_t \cdot \pi}{c_{\text{level2}} \left(\frac{\sqrt{2 \cdot n_r}}{L_{\text{PL2}}}\right)} \sqrt{\frac{r_{\text{end}}}{\left(r_{\text{end}} - 2 \cdot L_{\text{PL2}} \sqrt{2 \cdot n_r}\right)}}$$

$$\Rightarrow 4 \cdot n_r \equiv n_t \cdot \pi \sqrt{\frac{r_{\text{end}}}{\left(r_{\text{end}} - 2 \cdot L_{\text{PL2}} \sqrt{2 \cdot n_r}\right)}} \Rightarrow 16 \cdot n_r^2$$

$$\equiv n_t^2 \cdot \pi^2 \left(\frac{r_{\text{end}}}{\left(r_{\text{end}} - 2 \cdot L_{\text{PL2}} \sqrt{2 \cdot n_r}\right)}\right)$$

$$\Rightarrow r_{\text{end}} = \frac{2 \cdot L_{\text{PL2}} \sqrt{2 \cdot n_r}}{1 - \frac{n_t^2 \cdot \pi^2}{16 \cdot n_r^2}} = \frac{32 \cdot L_{\text{PL2}} \sqrt{2} \cdot n_r^{5/2}}{16 \cdot n_r^2 - n_t^2 \cdot \pi^2}. \tag{138}$$

The constraint behind (138) is to force the quantized time into a direct proportional connection with the equally quantized Schwarzschild radius via the connector c_{level2}.

Another and obviously more reasonable boundary condition can be formulated by using r_{end} at the position of r_s and we demand

$$\Rightarrow \frac{r_{\text{end}}}{c_{\text{level2}}} \equiv t_{n_t}(n_t, r_s) \Rightarrow \frac{r_{\text{end}}}{c_{\text{level2}}} \equiv \frac{n_t \cdot \pi}{c_{\text{level2}} \left(\frac{\sqrt{2 \cdot n_r}}{L_{\text{PL2}}}\right)} \sqrt{\frac{r_{\text{end}}}{\left(r_{\text{end}} - 2 \cdot L_{\text{PL2}} \sqrt{2 \cdot n_r}\right)}}$$

$$\Rightarrow r_{\text{end}}^2 \left(r_{\text{end}} - 2 \cdot L_{\text{PL2}} \sqrt{2 \cdot n_r}\right) \equiv \frac{n_t^2 \cdot \pi^2 L_{\text{PL2}}^2}{2 \cdot n_r} r_{\text{end}}$$

$$\Rightarrow r_{\text{end}} = L_{\text{PL2}} \left(\sqrt{2 \cdot n_r} + \sqrt{2 \cdot n_r + \frac{n_t^2 \cdot \pi^2}{2 \cdot n_r}}\right), \tag{139}$$

where we have only given the positive solution of the three possible r_{end} values.

For convenience, the equations above can alternatively be given with the Schwarzschild radius instead of the quantum number n_r. In this case, they do read as follows for $r_s / c_{\text{level2}} = t$:

$$\Rightarrow \frac{r_{s n_r}(n_r)}{c_{\text{level2}}} \equiv t_{n_t}(n_t, r_s) \Rightarrow \frac{r_{s n_r}(n_r)}{c_{\text{level2}}} \equiv \frac{n_t \cdot \pi}{c_{\text{level2}} \left(\frac{r_{s n_r}(n_r)}{2 \cdot L_{LP2}^2}\right)} \sqrt{\frac{r_{\text{end}}}{\left(r_{\text{end}} - r_{s n_r}(n_r)\right)}}$$

$$\Rightarrow \left[r_{s n_r}(n_r)\right]^4 \left(r_{\text{end}} - r_{s n_r}(n_r)\right) - 4 \cdot L_{LP2}^4 n_t^2 \cdot \pi^2 r_{\text{end}} \equiv 0, \tag{140}$$

$$r_{\text{end}} = \frac{r_{sn_r}(n_r)}{2} \cdot \left(1 + \sqrt{1 + \frac{n_t^2 \cdot \pi^2}{4 \cdot n_r^2}}\right) \tag{141}$$

and for $r_{\text{end}}/c_{\text{level2}} = t$:

$$\Rightarrow \frac{r_{\text{end}}}{c_{\text{level2}}} \equiv t_{n_t}(n_t, r_s) \Rightarrow \frac{r_{\text{end}}}{c_{\text{level2}}} \equiv \frac{n_t \cdot \pi}{c_{\text{level2}}\left(\frac{r_s}{2 \cdot L_{\text{PL2}}^2}\right)} \sqrt{\frac{r_{\text{end}}}{(r_{\text{end}} - r_s)}}$$

$$\Rightarrow r_{\text{end}} = \frac{1}{2}\left(r_s + \frac{\sqrt{16 \cdot L_{\text{PL2}}^4 n_t^2 \cdot \pi^2 + r_s^4}}{r_s}\right). \tag{142}$$

Discussion with Respect to $r_s(n_r)/t(n_t) \doteq c_{\text{level2}}$

At first we want to consider condition (138) respectively (140).

As an example, we solve the equation (140) for the Schwarzschild radius of the electron ($r_{se} = 1.35289*10^{-57}$ m) and obtain r_{end} to be always near r_{se} for smaller n. The same is found for bigger r_s because the equation above could be rewritten in the following form:

$$\Rightarrow r_{\text{end}} = \frac{r_{sn_r}(n_r)}{1 - \frac{4 \cdot L_{LP2}^4 n_t^2 \cdot \pi^2}{[r_{sn_r}(n_r)]^4}} \tag{143}$$

and the term $\frac{4 \cdot L_{LP2}^4 n_t^2 \cdot \pi^2}{[r_{sn_r}(n_r)]^4}$ is always extremely small for $n_t^2 \geq 1$ and $r_s > r_{se}$. Even the current upper bound for the mass of the smallest neutrino with about 2/5110 times the electron mass would give no significant deviation of r_s to r_{end}. It should be noted, however, that we always find a r_{end} to be slightly bigger than the Schwarzschild radius r_s. Thus, even for the object's "mass" (whatever this might be, because in our picture mass is just oscillations trapped in a certain region of space and time) being inside r_s the object's size would be slightly bigger than r_s. With bigger n_r, especially those needed for the r_s of massive elementary particles like the electron, we can almost continuously go to infinity for r_{end} with n_t being allowed to become a huge number before (143) switches to negative and thus impossible r_{end}. The situation is different at smaller n_r. Here only a certain number of n_t is allowed and there is no quasi-continuity as we found for bigger r_s respectively n_r. This automatically quantizes r_{end} rather severely for smaller n_r. We might see this as the quantization of space in the vicinity of a Schwarzschild object.

Interestingly, with the quasi-continuous behavior with respect to n_t for bigger n_r and the subsequent possibility for r_{end} to ALMOST become infinity, we find that for time steps t_n respectively n_t still just fulfilling the condition $16 \cdot n_r^2 > n_t^2 \cdot \pi^2$, the amount of space over which the Schwarzschild object respectively its oscillations stretch (as long as we intend to interpret r_{end} as such) can be huge.

Nevertheless, the setting $r_s(n_r)$ over $t(n_t)$ cannot be considered reasonable for a boundary condition, because it gives for all $t(n_t)$

$$\Rightarrow r(t) = c_{\text{level2}} \cdot t_{n_t} = 2 \cdot L_{\text{PL2}} \sqrt{2 \cdot n_r} = r_{sn_r}(n_r) \Rightarrow \frac{r_{sn_r}(n_r)}{t_{n_t}} = c_{\text{level2}},$$

$$(144)$$

which we directly have evaluated from equation (137).

Still, for completeness, we here evaluate a few results for the first n_r and n_t with respect to r_s and r_{end}. In the case of smallest possible n_r and subsequently allowed n_t (we demand the r_{end} stays positive) we obtain:

n_r & $n_t =\{\text{allowed}\}$	$n_r = 1$ & $n_t =\{1\}$	$n_r = 2$ & $n_t =\{1,2\}$	$n_r = 3$ & $n_t =\{1,2,3\}$	$n_r = 4$ & $n_t =\{1,2,3,4,5\}$
r_s in L_{PL2}	2.828	4	4.899	5.657
r_{end} in L_{PL2}	$\{7.382\}$	$\{4.729,$ $10.439\}$	$\{5.259,$ $6.749,$ $12.786\}$	$\{5.884, 6.689,$ $8.663\ 14.764,$ $156.39\}$

Discussion with Respect to $r_{\text{end}}(n_r, n_t)/t(n_r, n_t) = c_{\text{level2}}$

With the setting $r_{\text{end}}(n_r, n_t)/t(n_r, n_t) = c_{\text{level2}}$ or $r_{\text{end}}(r_s(n_r), n_t)/t(r_s(n_r), n_t) = c_{\text{level2}}$ we clearly have defined r_{end} to be the outermost 2-sphere in space at which the Schwarzschild object could be detected. As this parameter is quantized, we could consider also the space, in this case the r coordinate, being quantized.

The conceptual problem we face, however, lays in the fact the c_{level2} is not necessarily equal to the ordinary speed of light in vacuum, c. In order to make Schwarzschild radii possible, which are smaller than the Planck length, we had to demand the combined parameter $\frac{\hbar \cdot G}{c^3}$ to be much smaller than the classical constants would give. Instead, we had used the idea of a substructured Matryoschka-like space and applied new constants on the second sublevel with $\frac{\hbar_{\text{level2}} \cdot G_{\text{level2}}}{c_{\text{level2}}^3} = \frac{L_{\text{Planck}}^4}{m^2} = L_{\text{PL2}}^2$ (c.f. equation (133)). This was motivated by previous results [16, 23]. This might result in a much bigger speed of light for the level two. If this was the case, however, then the transfer of any information on level two would be processed with that speed, too. In order not to contradict this conclusion with the maximum speed c at our level (level 1), we have to find a process restricting this potentially higher speed of light of level two to scales below the Planck level. The same holds for the other two constants, of course (as long as they differ from their level 1 values).

How to Evaluate the Speed of Light of the Level Underneath?

The simplest explanation would be that the coordinates at level 2 are extremely curled up and that we, at level 1, only see an effective outcome of the structure resulting from the many times folded-up and or curved subspace. With L_{PL2} known, we can even evaluate how much the wiggling-up must have compactified the "inner" space. We must exactly have the number of N times the L_{PL2} sub-lengths curled up into one Planck length L_{Planck} in order to obtain the requested new level 2 constants. A structural suggestion was already made in [23] (see also [16]). There it was assumed that little sub-2 spheres ("Friedmanns") form on the surfaces of the Planck-sized elements of our space level. Thus, these Planck-sized objects ("Level 1 Friedmanns") are quasi 2-spheres, because the 2-spheres building their surfaces are extremely thin compared to the diameter of the Planck- or Level 1 Friedmanns. With the little level two Friedmanns (F2) having a radius of L_{PL2} and the Level 1 Friedmanns (F1) having a radius of L_{Planck}, the calculation is quite simple. At first we evaluate the number of F2s fitting in the surface of F1. Thereby we assume the F2s to sit so that we can imagine their projected surface (which is a circle of surface $\pi \cdot L_{\text{PL2}}^2$) distributed over the F1 surface such that we have $N = \frac{4 \cdot \pi \cdot L_{\text{Planck}}^2}{\pi \cdot L_{\text{PL2}}^2}$. This would not change much the total surface of the Planck-sized object (only 4 times the "smooth surface" value of $4 \cdot \pi \cdot L_{\text{Planck}}^2$). However, when evaluating a light signal traveling along the distance from pole to pole of an F1 object, the distance would become

$N \cdot \pi \cdot L_{PL2} = \frac{4 \cdot L_{Planck}^2}{L_{PL2}^2} \cdot \pi \cdot L_{PL2} = \frac{4 \cdot \pi \cdot L_{Planck}^2}{L_{PL2}}$. The observer in level 1 would see the signal only travel a short distance of 2 times L_{Planck} with the speed c in the time period T. In the subspace-level reality in the same time T, however, the signal has to travel a much larger distance and we obtain the speed c_{level2} of this propagation to be

$$\frac{N \cdot \pi \cdot L_{PL2}}{c_{level2}} = \frac{4 \cdot \pi \cdot L_{Planck}^2}{c_{level2} \cdot L_{PL2}} = \frac{2 \cdot L_{Planck}}{c}$$

$$\Rightarrow \quad c_{level2} = 2 \cdot \pi \cdot c \cdot \frac{L_{Planck}}{L_{PL2}} = 1.16546 \cdot 10^{44} \frac{m}{s}. \tag{145}$$

It should be noted that such a higher speed of light at the sublevel simplifies a few cosmological problems with respect to the homogeneity of space, because it reduces the need for dramatic inflation during, respectively right after the big bang. Furthermore, the suggested substructure might also explain the inflation itself as some kind of rapid moving of the F2 objects out of volume into the surfaces of the F1s. This, however, will be considered elsewhere.

Further, it should be noted that it is not unlikely that also the level two could be built from a sublevel and so force. We call this the Matryoschka character of space [22].

Conclusions to Quantized Schwarzschild

In this work, the author has applied the method of quantization the line element of an arbitrary space with smooth metric on the metric of a Schwarzschild object.

It was shown that this automatically quantizes time, mass (Schwarzschild radius) and the space around the Schwarzschild object.

It was also shown that gravitation, if based upon the Planck scale, would not allow for masses as small as known for the elementary particles like electron, positron or the quarks. The problem was solved by the assumption of a sublevel on which the gravitational force is built.

We evaluated the necessary scale parameters for this subspace thereby assuming a most simple Matryoschka structure of space.

We could also derive the sublevel speed of light to be about $3.88756*10^{35}$ times the classical speed of light c with $c = 299792458$ meters per second.

Chapter 5

Matter–Antimatter Asymmetry

Application to Dirac–Schwarzschild Particles at Rest

For simplicity, we will perform the following discussion on the simple metric of a Schwarzschild particle, but similar results are also obtained more complex metrics like Kerr, Kerr–de Sitter or Kerr–Newman and so on. In the classical Schwarzschild coordinates, we have the g^{00} compontent to be (with r_S giving the Schwarzschild radius of the particle in question):

$$g^{00} = c^{-2} \left(1 - \frac{r_S}{r} \right)^{-1} \tag{146}$$

and so the solutions for the generalized metric Dirac equation (see theory of the section "The Quantized Schwarzschild Metric" above) evolve to

$$\xi_0^{\text{total}} = x_0 + F\left(f\left(x^0 \right) \right) = x_0 + \left\{ \begin{array}{l} p_0 \; \cdot \; \left(C\,[1]_{\pm}^{\pm} \; \cdot \; e^{\pm x^0 \sqrt{1-\frac{r_S}{r}} \; \cdot \; (E \pm i \, \cdot \, \mu)} \right) \\ p_0 \; \cdot \; \left(C\,[2]_{\pm}^{\pm} \; \cdot \; e^{\pm i \, \cdot \, x^0 \sqrt{1-\frac{r_S}{r}} \; \cdot \; (E \pm \mu)} \right) \end{array} \right\}_{\Omega} \tag{147}$$

where we have sucked the c into the other constants (or set it to 1).

Now we could assume that the eigenstates E are somehow connected with the mass of the particles and thus, with their Schwarzschild radius r_S (c.f. equation (237)). The Schwarzschild radius itself, however, is mass times certain constants. As antimatter comes with "negative energy" respectively

The Theory of Everything: Quantum and Relativity is Everywhere – A Fermat Universe
Norbert Schwarzer
Copyright © 2020 Jenny Stanford Publishing Pte. Ltd.
ISBN 978-981-4774-47-5 (Hardcover), 978-1-315-09975-0 (eBook)
www.jennystanford.com

"negative mass" solutions we should consequently also assume this to holt true for the sign of the Schwarzschild radius within the metric g^{ij}. There is no need to bother about the Einstein tensor as the Einstein field equations will still be fulfilled (Einstein tensor and other things, like the Kretschmann scalar, will not be changed by a change of sign of r_S).

It seems to be a question of definition which one belongs to which, but apparently matter particles are connected with exponent-terms $\sqrt{1 + \frac{r_S}{r}}$, while the antimatter particles have exponents with $\sqrt{1 - \frac{r_S}{r}}$ (Here we applied a definition where the matter would win—one might call this cheating). This, of course, makes the effective mass in comparison with the classical Dirac solution a function of r, which is to say the distance to the particle. However, with the Schwarzschild radius usually being very small this should not become detectable in ordinary experimental setups and daily physical experiences. When annihilation of suitable pairs of such Dirac–Schwarzschild matter and antimatter particles takes place, however, the "effective mass asymmetry" should lead to dominant-recessive scenarios for the two agents.

The ratio of any matter–antimatter asymmetry η should be clearly determined by this "effective mass" asymmetry and we could write

$$1 + \eta = \frac{\sqrt{1 + \frac{r_S}{r}}}{\sqrt{1 - \frac{r_S}{r}}} = \sqrt{\frac{r + r_S}{r - r_S}}. \tag{148}$$

With the "universal annihilation" having led to a ratio known to be $\eta = (6.19 \pm 0.14) \times 10^{-10}$ (c.f. [24]) the process must have taken place when the Schwarzschild–Dirac particles were about $1{,}61551*10^9$ times their own Schwarzschild radii apart. In case annihilation had taken place later, much less matter would have survived. That this asymmetry has not contributed to a charged universe can easily be conclude from the Reissner–Nordström metric, wherein the charge Q of the particle, like the radius, comes as a square Q^2 [25, 26].

Chapter 6

Generalization of "The Recipe": From $\hbar$ to the Planck Tensor

Generalization to Non-diagonal Metrics

It is easy to prove that the scalar product of the following vector:

$$V_\Omega = \begin{cases} a+b+c+d+e, a+b+c+d-e, a+b+c-d+e, a+b+c-d-e, \\ a+i(b-c)+d+e, a+i(b-c)+d-e, a+i(b-c)-d+e, a+i(b-c)-d-e, \\ a-i(b-c)+d+e, a-i(b-c)+d-e, a-i(b-c)-d+e, a-i(b-c)-d-e, \\ a-b-c+d+e, a-b-c+d-e, a-b-c-d+e, a-b-c-d-e \end{cases}$$

$$\equiv [a \pm I(b \pm c) \pm d \pm e]_\Omega \tag{149}$$

gives

$$V_\Omega \cdot V_\Omega = C \cdot \left(a^2 + 2 \cdot b \cdot c + d^2 + e^2\right). \tag{150}$$

This allows us to treat metric shear components in the same easy way as we did with diagonal elements before. Again, we consider the intelligent zero of

The Theory of Everything: Quantum and Relativity is Everywhere – A Fermat Universe
Norbert Schwarzer
Copyright © 2020 Jenny Stanford Publishing Pte. Ltd.
ISBN 978-981-4774-47-5 (Hardcover), 978-1-315-09975-0 (eBook)
www.jennystanford.com

a line element:

$$ds^2_{\text{total}} - ds^2_{\text{total}} = \gamma^{nm} d\xi_n d\xi_m - \gamma_{nm} d\xi^n d\xi^m$$

$$= \gamma^{nm}(dx_n dx_m + dx_n d\,(p_m f_m) + d\,(p_n f_n)\,dx_m$$

$$+ d\,(p_n f_n)\,d\,(p_m f_m)) - \gamma_{nm}(dx^n dx^m + dy^n dx^m$$

$$+ dx^n dy^m + dy^n dy^m)$$

$$= \gamma^{nm}(dx_n dx_m + dx_n d\,(p_m f_m) + d\,(p_n f_n)\,dx_m$$

$$+ d\,(p_n f_n)\,d\,(p_m f_m)) - \gamma_{nm}$$

$$\left(dx^n dx^m + d\left(\gamma^{ni} \frac{\partial f}{\partial x^i} \right) dx^m + dx^n d\left(\gamma^{mi} \frac{\partial f}{\partial x^i} \right) \right.$$

$$\left. + d\left(\gamma^{ni} \frac{\partial f}{\partial x^i} \right) d\left(\gamma^{mi} \frac{\partial f}{\partial x^i} \right) \right)$$

$$\Rightarrow d\xi_n^{\text{total}} = dx_n^{\text{classic}} + dy_n^{\text{quantum}} = dx_n^{\text{classic}}$$

$$+ d\,(p_n f_n\,(x))^{\text{quantum}}$$

$$\Rightarrow \xi_n^{\text{total}} = x_n^{\text{classic}} + y_n^{\text{quantum}} = x_n^{\text{classic}} + p_n f_n\,(x)^{\text{quantum}}\,.$$

$$(151)$$

This time, however, we expand it as follows:

$$ds^2_{\text{total}} - ds^2_{\text{total}} = 0 = \gamma^{nm} d\xi_n d\xi_m - \gamma_{nm} d\xi^n d\xi^m$$

$$= \gamma^{nm}\,(dx_n dx_m + d\mathbf{x}_n h \cdot \mathbf{e}_m dy + h \cdot \mathbf{e}_n dy \cdot dx_m$$

$$+ \mathbf{e}_n \cdot \mathbf{e}_m h^2 dy^2) - \gamma_{nm}$$

$$\left(dx^n dx^m + d\left(\gamma^{ni} \frac{\partial h}{\partial x^i} \right) dx^m + dx^n d\left(\gamma^{mi} \frac{\partial h}{\partial x^i} \right) \right.$$

$$\left. + d\left(\gamma^{ni} \frac{\partial h}{\partial x^i} \right) d\left(\gamma^{mj} \frac{\partial h}{\partial x^j} \right) \right)$$

$$= \gamma^{nm}\,(dx_n dx_m + d\mathbf{x}_n \cdot \mathbf{f}_m dy + \mathbf{f}_n dy \cdot d\mathbf{x}_m + \mathbf{f}_n \cdot \mathbf{f}_m dy^2)$$

$$- \left(\gamma_{nm} dx^n dx^m + d\left(\gamma^{ni} \frac{\partial \mathbf{f}_n}{\partial x^i} \right) \cdot \mathbf{e}_m dx^m + d\left(\gamma^{mi} \frac{\partial \mathbf{f}_m}{\partial x^i} \right) \right.$$

$$\left. \times \mathbf{e}_n dx^n + d\left(\gamma^{ni} \frac{\partial \mathbf{f}_n}{\partial x^i} \right) d\left(\gamma^{mj} \frac{\partial \mathbf{f}_m}{\partial x^j} \right) \right)$$

$$\Rightarrow d\xi_n^{\text{total}} = d\mathbf{x}_n^{\text{classic}} + d\mathbf{y}_n^{\text{quantum}} = d\mathbf{x}_n^{\text{classic}} + d\,(\mathbf{e}_n h\,(x))^{\text{quantum}}$$

$$\Rightarrow \xi_n^{\text{total}} = \mathbf{x}_n^{\text{classic}} + \mathbf{y}_n^{\text{quantum}} = \mathbf{x}_n^{\text{classic}} + \mathbf{f}_n\,(x)^{\text{quantum}} \qquad (152)$$

where we have used the definition for base vectors and metrics and the subsequent properties:

$$h \cdot \mathbf{e}_m = f_m; \quad d\mathbf{x}_n \cdot d\mathbf{y}_m = d\mathbf{x}_n h \cdot \mathbf{e}_m dy$$

$$\gamma^{nm} = \mathbf{e}^n \cdot \mathbf{e}^m; \quad \gamma_{nm} = \mathbf{e}_n \cdot \mathbf{e}_m; \quad \mathbf{e}^m \cdot \mathbf{e}_n = \delta_n^m;$$

$$\gamma_{ni}\gamma^{im} = \delta_n^m; \quad \gamma^{nm} \cdot \mathbf{e}_n = \mathbf{e}^m; \quad \gamma_{nm} \cdot \mathbf{e}^m = \mathbf{e}_n. \tag{153}$$

In principle, four equations result from the intelligent zero given above:

$$ds_{\text{total}}^2 - ds_{\text{total}}^2 = 0$$

$$\Rightarrow \text{I.} \quad \gamma^{nm} dx_n dx_m - \gamma_{nm} dx^n dx^m = 0$$

$$\Rightarrow \text{II.} \quad \gamma^{nm} \mathbf{f}_n dy \cdot d\mathbf{x}_m - dy\left(\gamma^{ni}\frac{\partial \mathbf{f}_n}{\partial x^i}\right) \cdot \mathbf{e}_m dx^m = 0$$

$$\Rightarrow \text{III.} \quad \gamma^{nm} d\mathbf{x}_n \cdot \mathbf{f}_m dy - dy\left(\gamma^{mi}\frac{\partial \mathbf{f}_m}{\partial x^i}\right) \cdot \mathbf{e}_n dx^n = 0$$

$$\Rightarrow \text{IV.} \quad \gamma^{nm}\mathbf{f}_n \cdot \mathbf{f}_m dy^2 - \left(\left(\gamma^{ni}\frac{\partial \mathbf{f}_n}{\partial x^i}\right)\right)\left(\gamma^{mj}\frac{\partial \mathbf{f}_m}{\partial x^j}\right) dy^2 = 0. \tag{154}$$

While equation I is already (automatically) satisfied, we only have to watch equations II to IV. In order to keep things simple for the identical equations II and III, we have introduced the vector $d\mathbf{x}_n = \mathbf{e}_n dy$ with $\mathbf{e}^n d\mathbf{x}_n = \mathbf{e}^n \cdot \mathbf{e}_n \cdot dy = 4 \cdot dy$, giving us for equation II

$$\Rightarrow \text{II.} \quad 4f \cdot dy^2 - \left(\frac{\partial f}{\partial x^i}\right) dx^i dy = 0 = 4f \cdot dy - \left(\frac{\partial f}{\partial x^i}\right)\gamma^{ij} dx_j$$

$$= 4f \cdot dy - \left(\frac{\partial f}{\partial x^i}\right)\gamma^{ij}\mathbf{e}_j \cdot \hat{i}_j dy$$

$$\Rightarrow \text{III.} \quad 4f \cdot dy^2 - \left(\frac{\partial f}{\partial x^i}\right) dx^i dy = 0 = 4f \cdot dy - \left(\frac{\partial f}{\partial x^i}\right)\gamma^{ij}\mathbf{e}_j \cdot \hat{i}_j dy. \tag{155}$$

We find that for all equations we can extract factors in the following form (here only shown for IV where the factors are roots):

$$V_\Omega = C \cdot \left\{ \begin{array}{l} i \cdot \left(\pm\left(\sqrt{\frac{\gamma^{(ij)}}{2}}\mathbf{f}_{(i)} + \sqrt{\frac{\gamma^{(ij)}}{2}}\mathbf{f}_{(j)}\right)\right) \pm \left(\gamma^{(ij)}\frac{\partial \mathbf{f}_{(j)}}{\partial x^{(i)}} + \gamma^{(ij)}\frac{\partial \mathbf{f}_{(j)}}{\partial x^{(i)}}\right) \\ i \cdot \left(\pm i \cdot \left(\sqrt{\frac{\gamma^{(ij)}}{2}}\mathbf{f}_{(i)} + \sqrt{\frac{\gamma^{(ij)}}{2}}\mathbf{f}_{(j)}\right)\right) \pm i \cdot \left(\gamma^{(ij)}\frac{\partial \mathbf{f}_{(j)}}{\partial x^{(i)}} + \gamma^{(ij)}\frac{\partial \mathbf{f}_{(j)}}{\partial x^{(i)}}\right) \end{array} \right\}$$

$$\Rightarrow \text{IV.} \quad V_\Omega \cdot V_\Omega = \gamma^{mn}\mathbf{f}_n \cdot \mathbf{f}_m dy^2 - \left(\left(\gamma^{ni}\frac{\partial \mathbf{f}_n}{\partial x^i}\right)\right)\left(\gamma^{mj}\frac{\partial \mathbf{f}_m}{\partial x^j}\right) dy^2 = 0. \tag{156}$$

Here the notation (ij) means only the components of the tensor in question and (ii) stands for equal indices (diagonal elements) while (ij) stands for

$i \neq j$ (shear components). This gives us a set of Dirac-like equations still containing the metric, including shear components. In order to fulfill equation IV we can demand

$$V_\Omega = C \cdot \left\{ \begin{array}{l} i \cdot 2f \pm \sqrt{\gamma^{(ii)}} \frac{\partial f}{\partial x^{(i)}} \pm \left(\frac{\sqrt{\gamma^{(ij)}}}{2} \frac{\partial f}{\partial x^{(i)}} + \frac{\sqrt{\gamma^{(ij)}}}{2} \frac{\partial f}{\partial x^{(j)}} \right) \\[3mm] i \cdot 2f \pm \sqrt{\gamma^{(ii)}} \frac{\partial f}{\partial x^{(i)}} \pm i \cdot \left(\frac{\sqrt{\gamma^{(ij)}}}{2} \frac{\partial f}{\partial x^{(i)}} - \frac{\sqrt{\gamma^{(ij)}}}{2} \frac{\partial f}{\partial x^{(j)}} \right) \end{array} \right\}_\Omega = 0$$

(157)

for all its components, but it needs to be pointed out here, that again so called virtual parameters offer a certain degree of freedom. So, we find that also vectors V_Ω of the form below (158) would still satisfy equation IV.

$$V_\Omega = C \cdot \left\{ \begin{array}{l} i \cdot 2f \pm \sqrt{\gamma^{(ii)}} \frac{\partial f}{\partial x^{(i)}} \pm \left(\frac{\sqrt{\gamma^{(ij)}}}{2} \frac{\partial f}{\partial x^{(i)}} + \frac{\sqrt{\gamma^{(ij)}}}{2} \frac{\partial f}{\partial x^{(j)}} \right) \pm E \cdot f \\[3mm] i \cdot 2f \pm \sqrt{\gamma^{(ii)}} \frac{\partial f}{\partial x^{(i)}} \pm i \cdot \left(\frac{\sqrt{\gamma^{(ij)}}}{2} \frac{\partial f}{\partial x^{(i)}} - \frac{\sqrt{\gamma^{(ij)}}}{2} \frac{\partial f}{\partial x^{(j)}} \right) \pm E \cdot f \\[3mm] i \cdot 2f \pm \sqrt{\gamma^{(ii)}} \frac{\partial f}{\partial x^{(i)}} \pm \left(\frac{\sqrt{\gamma^{(ij)}}}{2} \frac{\partial f}{\partial x^{(i)}} + \frac{\sqrt{\gamma^{(ij)}}}{2} \frac{\partial f}{\partial x^{(j)}} \right) \pm i \cdot E \cdot f \\[3mm] i \cdot 2f \pm \sqrt{\gamma^{(ii)}} \frac{\partial f}{\partial x^{(i)}} \pm i \cdot \left(\frac{\sqrt{\gamma^{(ij)}}}{2} \frac{\partial f}{\partial x^{(i)}} - \frac{\sqrt{\gamma^{(ij)}}}{2} \frac{\partial f}{\partial x^{(j)}} \right) \pm i \cdot E \cdot f \end{array} \right\}_\Omega = 0.$$

(158)

This, however, gives us an eigenvalue equation of Dirac character with metric.

We will discuss the application of such a vector elsewhere in connection with particles [10].

Now we only need to show that the terms of II and III could also be constructed by the means of a scalar product of the vector V_Ω with another one B_Ω such that in the end we would obtain the equations II and III. With all elements in V_Ω demanded to be zero, also II and III would then automatically be fulfilled. We might find the requested by the means of a vector B_Ω:

$$B_\Omega \cdot V_\Omega = \sum_{i=1}^{\Omega} B_i \cdot V_i = [\text{II} \quad \text{or} \quad \text{III}].$$

(159)

Thus, we assume that it is always possible to find a vector B_Ω being a decompositions of dx_i and giving us the equations II and III via the scalar product $B_\Omega \cdot V_\Omega$ and with all elements of V_Ω being zero, II and III are also solved by the solutions of the root-factors of IV. The reader finds a more comprehensive discussion about the solutions in connection with II and III in reference [27].

Generalization of the "Clever Zero"

So far, we only used the simple expression (48) for our quantization procedure. In order also to consider other forms of clever zeros, it will be necessary to generalize our way to extract roots by the means of vectors (c.f. section "The Generalized Metric Dirac Operator"). Thus, we will try to extract the vector roots of expressions of the following form:

$$\Xi\left(ds_{q1}\right) - \Xi\left(ds_{q2}\right) = 0 = \Xi\left(e_i dx^i\right) - \Xi\left(e^i dx_i\right) \qquad (160)$$

with the line elements ds_{qn} $(n=1,2)$ and with general functions $\Xi[X]$.

The goal is to find a way to construct the expression above in a vector V_Ω form as follows:

$$\Xi\left(ds_{q1}\right) - \Xi\left(ds_{q2}\right) = \Xi\left(e_i dx^i\right) - \Xi\left(e^i dx_i\right) = \Xi\left(V_\Omega\left(ds_{q1}, ds_{q2}\right)\right). \qquad (161)$$

The Generalized "Vectorial Dirac Root"

Shall $n = \ln(\Xi[X])/\ln[X]$ and $n \geq 1$ $(n \, \varepsilon \, Z)$ then we find that the integral

$$
\int_{0-\varepsilon}^{2\pi-\varepsilon} \cdots \int_{0-\varepsilon}^{2\pi-\varepsilon} \frac{1}{k\cdot n} \sum_{j_1,\ldots,j_k=0}^{n-1} \left\{ \begin{array}{l} \left[\delta\left(\varphi_1 - \varphi_{1j}\right) + \ldots + \delta\left(\varphi_k - \varphi_{kj}\right)\right] \\ \times \Xi\left(a_1 + a_2 \cdot e^{i\varphi_1} + \ldots + a_{k+1} \cdot e^{i\varphi_k}\right) \end{array} \right\} d\varphi_1 \ldots d\varphi_k
$$

$$
= \frac{1}{k\cdot n} \sum_{j_1,\ldots,j_k=0}^{n-1} \Xi\left(a_1 + a_2 \cdot e^{i\cdot j_1\cdot\Delta\varphi_1} + \ldots + a_{k+1} \cdot e^{i\cdot j_k\cdot\Delta\varphi_k}\right)
$$

$$
= \Xi\left(a_1\right) + \ldots + \Xi\left(a_k\right); \quad \Delta\varphi_s = \frac{2\pi}{n} \qquad (162)
$$

gives the required result. The Greek delta thereby stands for the Dirac delta function. The vector V_Ω then automatically consist of all possible tuples of combinations of the j_k and reads

$$
V_\Omega = \Upsilon\left\{a_1 + a_2 \cdot e^{i\cdot j_1\cdot\Delta\varphi_1} + \ldots + a_{k+1} \cdot e^{i\cdot j_k\cdot\Delta\varphi_k}\right\}
$$

$$
= \left\{ \begin{array}{l} a_1 + a_2 + \ldots + a_{k+1} \\ a_1 + a_2 \cdot e^{i\cdot\Delta\varphi_1} + \ldots + a_{k+1} \\ \vdots \\ a_1 + a_2 \cdot e^{i\cdot j_1\cdot\Delta\varphi_1} + \ldots + a_{k+1} \cdot e^{i\cdot j_k\cdot\Delta\varphi_k} \\ \vdots \\ a_1 + a_2 \cdot e^{i\cdot(n-1)\cdot\Delta\varphi_1} + \ldots + a_{k+1} \cdot e^{i\cdot(n-1)\cdot\Delta\varphi_k} \end{array} \right\}. \qquad (163)
$$

It is relatively straightforward to prove that in the case of $n \; \varepsilon \; Q (Q \ldots$ rational numbers) (still $n > 1$) with $n = p/q$ the vector needs to be extended to

$$V_\Omega = \Upsilon \left\{ a_1 + a_2 \cdot e^{i \cdot j_1 \cdot \Delta\varphi_1} + \ldots + a_{k+1} \cdot e^{i \cdot j_k \cdot \Delta\varphi_k} \right\}$$

$$
= \left\{
\begin{array}{l}
a_1 + a_2 + \ldots + a_{k+1} \\[4pt]
a_1 + a_2 \cdot e^{i \cdot \Delta\varphi_1} + \ldots + a_{k+1} \\[4pt]
\vdots \\[4pt]
a_1 + a_2 \cdot e^{i \cdot j_1 \cdot \Delta\varphi_1} + \ldots + a_{k+1} \cdot e^{i \cdot j_k \cdot \Delta\varphi_k} \\[4pt]
\vdots \\[4pt]
a_1 + a_2 \cdot e^{i \cdot (n \cdot q - 1) \cdot \Delta\varphi_1} + \ldots + a_{k+1} \cdot e^{i \cdot (n \cdot q - 1) \cdot \Delta\varphi_k}
\end{array}
\right\} .
\qquad (164)
$$

In the integral above this simply leads to additional rotations of the phase angles as follows:

$$
\int_{0-\varepsilon}^{2\pi \cdot q - \varepsilon} \cdots \int_{0-\varepsilon}^{2\pi \cdot q - \varepsilon} \frac{1}{k \cdot n \cdot q} \sum_{j_1, \ldots, j_k = 0}^{n \cdot q - 1} \left\{ \begin{array}{l} \left[\delta \left(\varphi_1 - \varphi_{1j} \right) + \ldots + \delta \left(\varphi_k - \varphi_{kj} \right) \right] \\[4pt] \times \; \Xi \left(a_1 + a_2 \cdot e^{i \varphi_1} + \ldots + a_{k+1} \cdot e^{i \varphi_k} \right) \end{array} \right\} d\varphi_1 \ldots d\varphi_k
$$

$$
= \frac{1}{k \cdot n \cdot q} \sum_{j_1, \ldots, j_k = 0}^{n \cdot q - 1} \Xi \left(a_1 + a_2 \cdot e^{i \cdot j_1 \cdot \Delta\varphi_1} + \ldots + a_{k+1} \cdot e^{i \cdot j_k \cdot \Delta\varphi_k} \right)
$$

$$
= \Xi \left(a_1 \right) + \ldots + \Xi \left(a_k \right) ; \quad \Delta\varphi_s = \frac{2\pi}{n} .
\qquad (165)
$$

It is clear that in the case of real n with $n > 1$ the number of tuples becomes infinite and the integral evolves to a limes of the following form:

$$
\lim_{q \to \infty} \int_{0-\varepsilon}^{2\pi \cdot q - \varepsilon} \cdots \int_{0-\varepsilon}^{2\pi \cdot q - \varepsilon} \frac{1}{k \cdot n \cdot q} \sum_{j_1, \ldots, j_k = 0}^{n \cdot q - 1} \left\{ \begin{array}{l} \left[\delta \left(\varphi_1 - \varphi_{1j} \right) + \ldots + \delta \left(\varphi_k - \varphi_{kj} \right) \right] \\[4pt] \times \; \Xi \left(a_1 + a_2 \cdot e^{i \varphi_1} + \ldots + a_{k+1} \cdot e^{i \varphi_k} \right) \end{array} \right\} d\varphi_1 \ldots d\varphi_k
$$

$$
= \lim_{q \to \infty} \frac{1}{k \cdot n \cdot q} \sum_{j_1, \ldots, j_k = 0}^{n \cdot q - 1} \Xi \left(a_1 + a_2 \cdot e^{i \cdot j_1 \cdot \Delta\varphi_1} + \ldots + a_{k+1} \cdot e^{i \cdot j_k \cdot \Delta\varphi_k} \right)
$$

$$
= \Xi \left(a_1 \right) + \ldots + \Xi \left(a_k \right) ; \quad \Delta\varphi_s = \frac{2\pi}{n} .
\qquad (166)
$$

In the case of $0 > \ln(\Xi[X])/\ln[X] < 1$ we shall define a new $n = \ln(\Xi^{-1}[X])/\ln[X]$ with Ξ^{-1} giving the inverse function of f and then find that the integrals above have to be changed to

$$\Xi \left(\lim_{q\to\infty} \int_{0-\varepsilon}^{2\pi\cdot q-\varepsilon} \ldots \int_{0-\varepsilon}^{2\pi\cdot q-\varepsilon} \frac{1}{k\cdot n\cdot q} \sum_{j_1,\ldots,j_k=0}^{n\cdot q-1} \left\{ \begin{array}{l} [\delta(\varphi_1-\varphi_{1j})+\ldots+\delta(\varphi_k-\varphi_{kj})] \\ \times\, \Xi^{-1}\left(\Xi(a_1)+\Xi(a_2)\cdot e^{i\varphi_1}+\ldots+\Xi(a_{k+1})\cdot e^{i\varphi_k}\right) \end{array} \right\} d\varphi_1\ldots d\varphi_k \right)$$

$$= \Xi \left(\lim_{q\to\infty} \frac{1}{k\cdot n\cdot q} \sum_{j_1,\ldots,j_k=0}^{n\cdot q-1} \Xi^{-1}\left(\Xi(a_1)+\Xi(a_2)\cdot e^{i\cdot j_1\cdot\Delta\varphi_1}+\ldots+\Xi(a_{k+1})\cdot e^{i\cdot j_k\cdot\Delta\varphi_k}\right) \right)$$

$$= \Xi(a_1)+\ldots+\Xi(a_k); \quad \Delta\varphi_s = \frac{2\pi}{n}$$

(167)

which changes the vector to:

$$V_\Omega = \Upsilon \left\{ \Xi(a_1)+\Xi(a_2)\cdot e^{i\cdot j_1\cdot\Delta\varphi_1}+\ldots+\Xi(a_{k+1})\cdot e^{i\cdot j_k\cdot\Delta\varphi_k} \right\}$$

$$= \underbrace{\left\{ \begin{array}{l} \Xi(a_1)+\Xi(a_2)+\ldots+\Xi(a_{k+1}) \\ \Xi(a_1)+\Xi(a_2)\cdot e^{i\cdot\Delta\varphi_1}+\ldots+\Xi(a_{k+1}) \\ \ldots\; \Xi(a_1)+\Xi(a_2)\cdot e^{i\cdot j_1\cdot\Delta\varphi_1}+\ldots+\Xi(a_{k+1})\cdot e^{i\cdot j_k\cdot\Delta\varphi_k} \\ \ldots\; \Xi(a_1)+\Xi(a_2)\cdot e^{i\cdot(n\cdot q-1)\cdot\Delta\varphi_1}+\ldots+\Xi(a_{k+1})\cdot e^{i\cdot(n\cdot q-1)\cdot\Delta\varphi_k} \end{array} \right.}.$$

(168)

Now factorization via vectors of the V_Ω kind is possible for all sorts of functions Ξ. This also holds for $n < 0$ where the vector components then simply need to be inverted leading to either $(-1 > n > 0)$

$$V_\Omega = \Upsilon \left\{ \frac{1}{\Xi(a_1) + \Xi(a_2) \cdot e^{i \cdot j_1 \cdot \Delta\varphi_1} + \ldots + \Xi(a_{k+1}) \cdot e^{i \cdot j_k \cdot \Delta\varphi_k}} \right\} \qquad (169)$$

or $(n < -1)$:

$$V_\Omega = \Upsilon \left\{ \frac{1}{a_1 + a_2 \cdot e^{i \cdot j_1 \cdot \Delta\varphi_1} + \ldots + a_{k+1} \cdot e^{i \cdot j_k \cdot \Delta\varphi_k}} \right\}. \qquad (170)$$

In the following, we shall understand (168) or its extension to either $(-1 > n > 0)$ (169) or $(n < -1)$ (170) in its tuple-component structure using the simplified writing (here only given for the (168)-case)

$$
\begin{aligned}
V_\Omega &\equiv V_\Omega(a_1, a_2, \ldots, a_{k+1}) \\
&\equiv \Upsilon \left\{ \Xi(a_1) + \Xi(a_2) \cdot e^{i \cdot j_1 \cdot \Delta\varphi_1} + \ldots + \Xi(a_{k+1}) \cdot e^{i \cdot j_k \cdot \Delta\varphi_k} \right\} \\
&= \left\{
\begin{array}{l}
\Xi(a_1) + \Xi(a_2) + \ldots + \Xi(a_{k+1}) \\
\Xi(a_1) + \Xi(a_2) \cdot e^{i \cdot \Delta\varphi_1} + \ldots + \Xi(a_{k+1}) \\
\vdots \\
\Xi(a_1) + \Xi(a_2) \cdot e^{i \cdot j_1 \cdot \Delta\varphi_1} + \ldots + \Xi(a_{k+1}) \cdot e^{i \cdot j_k \cdot \Delta\varphi_k} \\
\vdots \\
\Xi(a_1) + \Xi(a_2) \cdot e^{i \cdot (n \cdot q - 1) \cdot \Delta\varphi_1} + \ldots + \Xi(a_{k+1}) \cdot e^{i \cdot (n \cdot q - 1) \cdot \Delta\varphi_k}
\end{array}
\right\}.
\end{aligned}
\qquad (171)
$$

And subsequently in similar form for the cases $(-1 > n > 0)$ (169) or $(n < -1)$ (170).

Now, especially in connection with certain metrics, which are containing non-diagonal (shear) components we might need to express vectors having the following properties:

$$\Xi(V_\Omega(a_1 + b_1, a_2)) - \Xi(V_\Omega(a_1, b_1, a_2)) = \Xi\left(\tilde{V}_\Omega(a_1, b_1, a_2)\right). \qquad (172)$$

We find the necessary structure by the means of

$$\tilde{V}_\Omega \equiv V_\Omega\left(\frac{a_1 + b_1}{\nu}, \frac{a_1}{\nu} \cdot \Xi^{-1}(-1), \frac{b_1}{\nu} \cdot \Xi^{-1}(-1), c \right), \qquad (173)$$

with ν being a constant which has to be calculated via $\nu = \Xi^{-1}(2)$. An application is given with the Kerr metrics (Kerr, Kerr–Newman and

Kerr–Newman–de Sitter) in [10] as a possible underlying metric for certain particles.

Examples for other "Vectorial Dirac Roots"

Simple square root with shear component with $\Xi(X) = X^2$

In our first example, we are looking for a vector V_Ω giving us the following scalar product:

$$
V_\Omega \cdot V_\Omega = V_\Omega
\begin{pmatrix}
\dfrac{a+b}{v}, \dfrac{a}{v} \cdot \Xi^{-1}(-1), \\[2mm]
\dfrac{b}{v} \cdot \Xi^{-1}(-1), c
\end{pmatrix}
\cdot V_\Omega
\begin{pmatrix}
\dfrac{a+b}{v}, \dfrac{a}{v} \cdot \Xi^{-1}(-1), \\[2mm]
\dfrac{b}{v} \cdot \Xi^{-1}(-1), c
\end{pmatrix}
$$

$$
= a \cdot b + c^2. \tag{174}
$$

With the knowledge gained in the section above, we can easily construct the necessary vector:

$$
V_\Omega \equiv V_\Omega \left(\frac{a+b}{4}, \frac{a}{4} \cdot \sqrt{-1}, \frac{b}{4} \cdot \sqrt{-1}, \frac{c}{2\sqrt{2}} \right)
$$

$$
= V_\Omega \left(\alpha \equiv \frac{a+b}{4}, \beta \equiv \frac{a}{4} \cdot i, \gamma \equiv \frac{b}{4} \cdot i, \delta \equiv \frac{c}{2\sqrt{2}} \right)
$$

$$
= \left\{
\begin{array}{l}
\alpha + \beta + \gamma + \delta, \alpha + \beta + \gamma - \delta, \alpha + \beta - \gamma + \delta, \alpha + \beta - \gamma - \delta, \\
\alpha - \beta + \gamma + \delta, \alpha - \beta + \gamma - \delta, \alpha - \beta - \gamma + \delta, \alpha - \beta - \gamma - \delta
\end{array}
\right\}.
$$

$$
\tag{175}
$$

We leave it to the reader to prove that this will give the required results as demanded in (174).

Simple square root with shear component with $\Xi(X) = X^2$ with virtual parameters E_i of various orders of "virtuality"

Now, we are looking for a vector V_Ω giving us the following scalar product:

$$
V_\Omega \cdot V_\Omega = V_\Omega
\begin{pmatrix}
\dfrac{a+b}{v}, \dfrac{a}{v} \cdot \Xi^{-1}(-1), \dfrac{b}{v} \cdot \Xi^{-1}(-1), \\[2mm]
c, E_1, E_2, \dots, E_j, \dots
\end{pmatrix}
$$

$$
\times V_\Omega
\begin{pmatrix}
\dfrac{a+b}{v}, \dfrac{a}{v} \cdot \Xi^{-1}(-1), \dfrac{b}{v} \cdot \Xi^{-1}(-1), \\[2mm]
c, E_1, E_2, \dots, E_j, \dots
\end{pmatrix}
= a \cdot b + c^2.
$$

$$
\tag{176}
$$

With the knowledge gained in the section above, we can easily construct the necessary vector:

$$V_\Omega \equiv V_\Omega \begin{pmatrix} \frac{a+b}{4}, \frac{a}{4} \cdot \sqrt{-1}, \frac{b}{4} \cdot \sqrt{-1}, \frac{c}{2\sqrt{2}}, E_1, e^{\frac{i\pi}{2}} E_1, E_2, e^{\frac{i\pi}{3}} E_2, e^{\frac{2i\pi}{3}} E_2, \\ E_3, e^{\frac{i\pi}{4}} E_3, e^{\frac{i\pi}{2}} E_3, e^{\frac{3i\pi}{4}} E_3, \ldots, E_j, e^{\frac{i\pi}{2 \cdot j}} E_j, e^{\frac{2i\pi}{2 \cdot j}} E_j, \ldots, e^{\frac{(2j-1)i\pi}{2 \cdot j}} E_j, \ldots \end{pmatrix}$$

$$= V_\Omega \left(\alpha \equiv \frac{a+b}{4}, \beta \equiv \frac{a}{4} \cdot i, \gamma \equiv \frac{b}{4} \cdot i, \delta \equiv \frac{c}{2\sqrt{2}}, E_1, \ldots \right)$$

$$= \left\{ \begin{array}{l} \alpha + \beta + \gamma + \delta \pm E_1 \pm e^{\frac{i\pi}{2}} E_1 \pm \ldots \pm e^{\frac{(2j-1)i\pi}{2 \cdot j}} E_j \pm \ldots, \\ \alpha + \beta + \gamma - \delta \pm E_1 \pm e^{\frac{i\pi}{2}} E_1 \pm \ldots \pm e^{\frac{(2j-1)i\pi}{2 \cdot j}} E_j \pm \ldots, \\ \alpha + \beta - \gamma + \delta \pm E_1 \pm e^{\frac{i\pi}{2}} E_1 \pm \ldots \pm e^{\frac{(2j-1)i\pi}{2 \cdot j}} E_j \pm \ldots, \\ \alpha + \beta - \gamma - \delta \pm E_1 \pm e^{\frac{i\pi}{2}} E_1 \pm \ldots \pm e^{\frac{(2j-1)i\pi}{2 \cdot j}} E_j \pm \ldots, \\ \alpha - \beta + \gamma + \delta \pm E_1 \pm e^{\frac{i\pi}{2}} E_1 \pm \ldots \pm e^{\frac{(2j-1)i\pi}{2 \cdot j}} E_j \pm \ldots, \\ \alpha - \beta + \gamma - \delta \pm E_1 \pm e^{\frac{i\pi}{2}} E_1 \pm \ldots \pm e^{\frac{(2j-1)i\pi}{2 \cdot j}} E_j \pm \ldots, \\ \alpha - \beta - \gamma + \delta \pm E_1 \pm e^{\frac{i\pi}{2}} E_1 \pm \ldots \pm e^{\frac{(2j-1)i\pi}{2 \cdot j}} E_j \pm \ldots, \\ \alpha - \beta - \gamma - \delta \pm E_1 \pm e^{\frac{i\pi}{2}} E_1 \pm \ldots \pm e^{\frac{(2j-1)i\pi}{2 \cdot j}} E_j \pm \ldots \end{array} \right\}. \tag{177}$$

Here the $\pm -$ sign stands for all tuples of variations of "+" and "−." We leave it to the reader to prove that this will give the required results as demanded in (174). In connection with (158) this would result in a system of partial differential equations of first order with an arbitrary number of eigenvalues of increasing order. An application of rather simple character is given in [10].

Simple cubic root $\Xi(X) = X^3$

We find that the vector

$$V_\Omega \equiv V_\Omega (a, b, c)$$

$$= \frac{1}{3} \left\{ \begin{array}{l} a + c + b, a + b + e^{\frac{2i\pi}{3}} c, a + b + e^{\frac{2 \cdot 2i\pi}{3}} c, \\ a + e^{\frac{2i\pi}{3}} b + c, a + e^{\frac{2i\pi}{3}} b + e^{\frac{2i\pi}{3}} c, a + e^{\frac{2i\pi}{3}} b + e^{\frac{2 \cdot 2i\pi}{3}} c, \\ a + e^{-\frac{2i\pi}{3}} b + c, a + e^{-\frac{2i\pi}{3}} b + e^{\frac{2i\pi}{3}} c, a + e^{-\frac{2i\pi}{3}} b + e^{\frac{2 \cdot 2i\pi}{3}} c \end{array} \right\} \tag{178}$$

gives us

$$\Xi [V_\Omega] = \Xi [V_\Omega (a, b, c)] = (V_\Omega (a, b, c))^3$$

$$= \sum_{i=1}^{\forall \Omega} [V_i]^3$$

$$= a^3 + b^3 + c^3. \tag{179}$$

Simple cubic root $\Xi(X) = X^3$ with virtual parameter c

We find that the vector

$$V_\Omega \equiv V_\Omega(a, b, c)$$

$$= \frac{1}{\sqrt[3]{18}} \left\{ \begin{array}{l} a + c + b, a + b + e^{\frac{2\mathbf{1}\pi}{6}} c, a + b + e^{\frac{2*2\mathbf{1}\pi}{6}} c, a + b + e^{\frac{3*2\mathbf{1}\pi}{6}} c, a + b + e^{\frac{4*2\mathbf{1}\pi}{6}} c, \\[4pt] a + b + e^{\frac{5*2\mathbf{1}\pi}{6}} c, a + e^{\frac{2\mathbf{1}\pi}{3}} b + c, a + e^{\frac{2\mathbf{1}\pi}{3}} b + e^{\frac{2\mathbf{1}\pi}{6}} c, a + e^{\frac{2\mathbf{1}\pi}{3}} b + e^{\frac{2*2\mathbf{1}\pi}{6}} c, \\[4pt] a + e^{\frac{2\mathbf{1}\pi}{3}} b + e^{\frac{3*2\mathbf{1}\pi}{6}} c, a + e^{\frac{2\mathbf{1}\pi}{3}} b + e^{\frac{4*2\mathbf{1}\pi}{6}} c, a + e^{\frac{2\mathbf{1}\pi}{3}} b + e^{\frac{5*2\mathbf{1}\pi}{6}} c, \\[4pt] a + e^{-\frac{2\mathbf{1}\pi}{3}} b + c, a + e^{-\frac{2\mathbf{1}\pi}{3}} b + e^{\frac{2\mathbf{1}\pi}{6}} c, a + e^{-\frac{2\mathbf{1}\pi}{3}} b + e^{\frac{2*2\mathbf{1}\pi}{6}} c, \\[4pt] a + e^{-\frac{2\mathbf{1}\pi}{3}} b + e^{\frac{3*2\mathbf{1}\pi}{6}} c, a + e^{-\frac{2\mathbf{1}\pi}{3}} b + e^{\frac{4*2\mathbf{1}\pi}{6}} c, a + e^{-\frac{2\mathbf{1}\pi}{3}} b + e^{\frac{5*2\mathbf{1}\pi}{6}} c \end{array} \right. \tag{180}$$

gives us

$$\Xi[V_\Omega] = \Xi[V_\Omega(a, b, c)] = (V_\Omega(a, b, c))^3 = \sum_{i=1}^{\Omega A} [V_i]^3 = a^3 + b^3. \tag{181}$$

Simple quartic root $\Xi(X) = X^4$

We find that the vector

$$V_\Omega \equiv V_\Omega(a, b, c)$$

$$= \frac{1}{4} \left\{ \begin{array}{l} a + b + c, a + b + e^{\frac{2\mathbf{1}\pi}{4}} c, a + b - e^{\frac{2*2\mathbf{1}\pi}{4}} c, a + b + e^{\frac{3*2\mathbf{1}\pi}{4}} c, \\[4pt] a + e^{\frac{2\mathbf{1}\pi}{4}} b + c, a + e^{\frac{2\mathbf{1}\pi}{4}} b + e^{\frac{2\mathbf{1}\pi}{4}} c, a + e^{\frac{2\mathbf{1}\pi}{4}} b + e^{\frac{2*2\mathbf{1}\pi}{4}} c, a + e^{\frac{2\mathbf{1}\pi}{4}} b + e^{\frac{3*2\mathbf{1}\pi}{4}} c, \\[4pt] a + e^{\mathbf{1}\pi} b + c, a + e^{\mathbf{1}\pi} b + e^{\frac{2\mathbf{1}\pi}{4}} c, a + e^{\mathbf{1}\pi} b + e^{\frac{2*2\mathbf{1}\pi}{4}} c, a + e^{\mathbf{1}\pi} b + e^{\frac{3*2\mathbf{1}\pi}{4}} c, \\[4pt] a + e^{\frac{3*2\mathbf{1}\pi}{4}} b + c, a + e^{\frac{3*2\mathbf{1}\pi}{4}} b + e^{\frac{2\mathbf{1}\pi}{4}} c, a + e^{\frac{3*2\mathbf{1}\pi}{4}} b + e^{\frac{2*2\mathbf{1}\pi}{4}} c, a + e^{\frac{3*2\mathbf{1}\pi}{4}} b + e^{\frac{3*2\mathbf{1}\pi}{4}} c \end{array} \right. \tag{182}$$

gives us

$$\Xi[V_\Omega] = \Xi[V_\Omega(a,b,c)] = (V_\Omega(a,b,c))^4 = \sum_{i=1}^{\Omega}[V_i]^4 = a^4 + b^4 + c^4. \tag{183}$$

Simple quartic root $\Xi(X) = X^4$ with virtual parameter c

We find that the vector

$$V_\Omega \equiv V_\Omega(a,b,c)$$

$$= \frac{1}{\sqrt[4]{32}} \left\{
\begin{aligned}
&a+b+c,\; a+b+e^{\frac{2i\pi}{8}}c,\; a+b+e^{\frac{2i\pi}{4}}c,\; a+b+e^{\frac{3*2i\pi}{8}}c,\; a+b+e^{\frac{2*2i\pi}{4}}c,\\
&a+b+e^{\frac{5*2i\pi}{8}}c,\; a+b+e^{\frac{3*2i\pi}{4}}c,\; a+b+e^{\frac{7*2i\pi}{8}}c,\\
&a+e^{\frac{2i\pi}{4}}b+c,\; a+e^{\frac{2i\pi}{4}}b+e^{\frac{2i\pi}{8}}c,\; a+e^{\frac{2i\pi}{4}}b+e^{\frac{2i\pi}{4}}c,\\
&a+e^{\frac{2i\pi}{4}}b+e^{\frac{3*2i\pi}{8}}c,\; a+e^{\frac{2i\pi}{4}}b+e^{\frac{2*2i\pi}{4}}c,\\
&a+e^{\frac{2i\pi}{4}}b+e^{\frac{5*2i\pi}{8}}c,\; a+e^{\frac{2i\pi}{4}}b+e^{\frac{3*2i\pi}{4}}c,\; a+e^{\frac{2i\pi}{4}}b+e^{\frac{7*2i\pi}{8}}c,\\
&a+e^{i\pi}b+c,\; a+e^{i\pi}b+e^{\frac{2i\pi}{8}}c,\; a+e^{i\pi}b+e^{\frac{2i\pi}{4}}c,\; a+e^{i\pi}b+e^{\frac{3*2i\pi}{8}}c,\\
&a+e^{i\pi}b+e^{\frac{2*2i\pi}{4}}c,\; a+e^{i\pi}b+e^{\frac{5*2i\pi}{8}}c,\; a+e^{i\pi}b+e^{\frac{3*2i\pi}{4}}c,\; a+e^{i\pi}b+e^{\frac{7*2i\pi}{8}}c,\\
&a+e^{\frac{3*2i\pi}{4}}b+c,\; a+e^{\frac{3*2i\pi}{4}}b+e^{\frac{2i\pi}{8}}c,\; a+e^{\frac{3*2i\pi}{4}}b+e^{\frac{2i\pi}{4}}c,\\
&a+e^{\frac{3*2i\pi}{4}}b+e^{\frac{3*2i\pi}{8}}c,\; a+e^{\frac{3*2i\pi}{4}}b+e^{\frac{2*2i\pi}{4}}c,\\
&a+e^{\frac{3*2i\pi}{4}}b+e^{\frac{5*2i\pi}{8}}c,\; a+e^{\frac{3*2i\pi}{4}}b+e^{\frac{7*2i\pi}{8}}c
\end{aligned}
\right\} \tag{184}$$

gives us

$$\Xi\left[V_\Omega\right] = \Xi\left[V_\Omega\left(a, b, c\right)\right] = \left(V_\Omega\left(a, b, c\right)\right)^4 = \sum_{i=1}^{\forall\Omega}\left[V_i\right]^4 = a^4 + b^4. \quad (185)$$

Now as we have seen how to factorize also more complicated general expressions $\Xi(X)$ instead of only $\Xi = X^2$ as so far applied in [7-17], we can move on to further generalizations with respect to our "clever zero line element quantization approach."

Extension/Generalization to Arbitrary Functional Approaches for $K(f_n)$

The Planck functional

For the following extensions we will concentrate only on simple Ξ-functions of the kind ^{2}. This goes without loss of generality and is only for the reason of simplicity and brevity. Also for the reason of simplicity, we will avoid distinguishing between the bold vectors dx_m, dy_m, $d\xi_m$, f_m and scalars. Meaning, that it goes without saying that these symbols have to be understood as dx_m, dy_m, $d\xi_m$, f_m.

As already hinted in [15] it is possible to use a functional setting $K(f)$ for the function f in (52):

$$ds_{\text{total}}^2 - ds_{\text{total}}^2 = \gamma^{nm}d\xi_n d\xi_m - \gamma_{nm}d\xi^n d\xi^m$$
$$= \gamma^{nm}\left(dx_n dx_m + dx_n dK\left(f_m\right) + dK\left(f_n\right)dx_m + dK\left(f_n\right)dK\left(f_m\right)\right)$$
$$-\gamma_{nm}\left(dx^n dx^m + dy^n dx^m + dx^n dy^m + dy^n dy^m\right)$$
$$= \gamma^{nm}\left(dx_n dx_m + dx_n dK\left(f_m\right) + dK\left(f_n\right)dx_m + dK\left(f_n\right)dK\left(f_m\right)\right)$$
$$-\gamma_{nm}\left(dx^n dx^m + d\left(\gamma^{ni}\frac{\partial f}{\partial x^i}\right)dx^m + dx^n d\left(\gamma^{mi}\frac{\partial f}{\partial x^i}\right)\right.$$
$$\left. +d\left(\gamma^{ni}\frac{\partial f}{\partial x^i}\right)d\left(\gamma^{mi}\frac{\partial f}{\partial x^i}\right)\right)$$

$$\text{with}\quad K\left(f_m\right) \equiv \left(\hat{F}\left(f_m\right)f_m\right)$$
$$\Rightarrow d\xi_n^{\text{total}} = dx_n^{\text{classic}} + dy_n^{\text{quantum}} = dx_n^{\text{classic}} + dK\left(f_n\left(x\right)\right)^{\text{quantum}}$$
$$\Rightarrow \xi_n^{\text{total}} = x_n^{\text{classic}} + y_n^{\text{quantum}} = x_n^{\text{classic}} + K\left(f_n\left(x\right)\right)^{\text{quantum}}.$$

$$(186)$$

Again, as an example, in the case of the Higgs field, we would have to apply the well-known function

$$K\left[f_m\right] = \hat{F}\left[f_m\right]f_m = \left(\mu \pm i \cdot \lambda \cdot f_m\right)f_m, \quad (187)$$

where we have used the usual quantum mechanical operator sign "∧" for the functional $\hat{F}\,[f_m]$. From our previous comparison with the classical Minkowski-based Klein–Gordon equation (c.f. (33)) we could also interpret the functional setting $K[f]$ as a Planck function instead of a Planck constant. From (30) we have

$$\hat{F}^2 = \frac{M^2 c^2}{\hbar^2} \quad \Rightarrow \quad K(f) = \underbrace{\hat{F}(f)f = p \cdot f}_{\text{classic Planck}} = \frac{M \cdot c}{\hbar} \cdot f. \tag{188}$$

Thus, interpreting "$p^* \ldots$" as the most simple K functional, one could understand the generalization of K as functional generalization of the Planck constant as follows:

$$\hat{F}^2 = \frac{M^2 c^2}{\hbar^2} \quad \Rightarrow \quad K(f) = \hat{F}(f)f = M \cdot c \cdot \hat{\hbar}(f) \cdot f \tag{189}$$

with $\hat{\hbar}(f)$ giving the functional Planck connection and becoming $\hat{\hbar}(f) \Rightarrow \hbar^{-1}$ in the simple classical case.

Extension/Generalization to Arbitrary Derivative Approaches: The Generalized Gradient of f_n

$$ds_{\text{total}}^2 - ds_{\text{total}}^2 = \gamma^{nm} d\xi_n d\xi_m - \gamma_{nm} d\xi^n d\xi^m$$

$$= \gamma^{nm} \left(dx_n dx_m + dx_n dK\,(f_m) + dK\,(f_n)\,dx_m + dK\,(f_n)\,dK\,(f_m) \right)$$

$$-\gamma_{nm} \begin{pmatrix} dx^n dx^m + d\,(D^n f)\,dx^m \\ + dx^n d\,(D^m f) + d\,(D^n f)\,d\,(D^m f) \end{pmatrix}$$

$$\text{with} \quad D^n f = D_1 \cdot \gamma^{ni} \frac{\partial f}{\partial x^i} + D_2 \cdot \gamma^{ni} e^j \frac{\partial}{\partial x^j} \frac{\partial f}{\partial x^i}$$

$$+ D_3 \cdot \gamma^{ni} e^j e^k \frac{\partial}{\partial x^k} \frac{\partial}{\partial x^j} \frac{\partial f}{\partial x^i} + \ldots \tag{190}$$

This is leading to a more complex functional approach for the "gradient setting" within the V_Ω vectors:

$$\widehat{V}_\Omega^m f_m = \left\{ D^m \pm i \cdot \sqrt{g^{mi}} \cdot e_i \cdot \hat{F} \right\}_\Omega f_m$$

$$= \left\{ D_1 \cdot \gamma^{mi} \frac{\partial f}{\partial x^i} \pm D_2 \cdot \gamma^{mi} e^j \frac{\partial}{\partial x^j} \frac{\partial f}{\partial x^i} \right.$$

$$\left. \pm D_3 \cdot \gamma^m e^j e^k \frac{\partial}{\partial x^k} \frac{\partial}{\partial x^j} \frac{\partial f}{\partial x^i} \pm \cdots \pm i \cdot \sqrt{g^{mi}} \cdot e_i \cdot \hat{F} \right\}_\Omega f_m \tag{191}$$

where, for the reason of simplicity, we have again assumed a simple Ξ-function of the kind ∧2.

Extension/Generalization to Higher-Order Planck Tensors

$$ds_{total}^2 - ds_{total}^2 = \gamma^{nm} d\xi_n d\xi_m - \gamma_{nm} d\xi^n d\xi^m$$

$$= \gamma^{nm} \left(dx_n dx_m + dx_n dK(f_m) + dK(f_n) dx_m + dK(f_n) dK(f_m) \right)$$

$$- \gamma_{nm} \left(dx^n dx^m + d\left(\gamma^{ni} \frac{\partial f}{\partial x^i} \right) dx^m + dx^n d\left(\gamma^{mi} \frac{\partial f}{\partial x^i} \right) \right.$$

$$\left. + d\left(\gamma^{ni} \frac{\partial f}{\partial x^i} \right) d\left(\gamma^{mi} \frac{\partial f}{\partial x^i} \right) \right)$$

$$= \gamma^{nm} \left(dx_n dx_m + dx_n dK(f_m) + dK(f_n) dx_m + dK(f_n) dK(f_m) \right)$$

$$- \left(\gamma_{nm} dx^n dx^m + \gamma_{nm} d\left(\gamma^{ni} \frac{\partial f}{\partial x^i} \right) dx^m + \gamma_{nm} dx^n d\left(\gamma^{mi} \frac{\partial f}{\partial x^i} \right) \right.$$

$$\left. + e_m e_n d\left(\gamma^{ni} \frac{\partial f}{\partial x^i} \right) d\left(\gamma^{mi} \frac{\partial f}{\partial x^i} \right) \right)$$

$$= \gamma^{nm} \left(dx_n dx_m + dx_n dK(f_m) + dK(f_n) dx_m + dK(f_n) dK(f_m) \right)$$

$$- \left(\gamma_{nm} dx^n dx^m + e_m d\left(\gamma^{ni} \frac{\partial f_n}{\partial x^i} \right) dx^m + e_m dx^n d\left(\gamma^{mi} \frac{\partial f_m}{\partial x^i} \right) \right.$$

$$\left. + d\left(\gamma^{ni} \frac{\partial f_n}{\partial x^i} \right) d\left(\gamma^{mi} \frac{\partial f_m}{\partial x^i} \right) \right)$$

with $\quad K(f_m) \equiv \tilde{K}\left(p \cdot \Phi_m + p^n \cdot \Phi_{mn} + p^{nk} \cdot \Phi_{mnk} + p^{nkl} \cdot \Phi_{mnkl} + \dots \right)$

$$\gamma^{mi} \frac{\partial f_m}{\partial x^i} \equiv \gamma^{mi} \frac{\partial}{\partial x^i} \left[\Phi_m + e^n \cdot \Phi_{mn} + \gamma^{nk} \cdot \Phi_{mnk} + e^l \gamma^{nk} \cdot \Phi_{mnkl} + \dots \right]$$

$$\Rightarrow d\xi_n^{\text{total}} = \overset{\text{classic}}{dx_n} + \overset{\text{quantum}}{dy_n} = \overset{\text{classic}}{dx_n} + \overset{\text{quantum}}{dK(f_n(x))}$$

$$\Rightarrow \overset{\text{total}}{\xi_n} = \overset{\text{classic}}{x_n} + \overset{\text{quantum}}{y_n} = \overset{\text{classic}}{x_n}$$

$$+ \overbrace{\tilde{K}\left(p \cdot \Phi_m + p^n \cdot \Phi_{mn} + p^{nk} \cdot \Phi_{mnk} + p^{nkl} \cdot \Phi_{mnkl} + \dots \right)}^{\text{quantum}}.$$

$$(192)$$

From (188) and our comparison with the Minkowski-like Klein–Gordon equation we know that the functional K is directly connected with the Planck constant $\hbar$, because we have

$$\hat{F}^2 = \frac{M^2 c^2}{\hbar^2} \quad \Rightarrow \quad K(f) = \underbrace{\hat{F}(f)f}_{\text{classic Planck}} = p \cdot f = \frac{M \cdot c}{\hbar} \cdot f. \qquad (193)$$

Applying the simplest K-Form to (192) by writing

$$K\left(f_m\right) \equiv p \cdot \Phi_m + p^n \cdot \Phi_{mn} + p^{nk} \cdot \Phi_{mnk} + p^{nkl} \cdot \Phi_{mnkl} + \dots$$

$$\gamma^{mi} \frac{\partial f_m}{\partial x^i} \equiv \gamma^{mi} \frac{\partial}{\partial x^i} \left[\Phi_m + e^n \cdot \Phi_{mn} + \gamma^{nk} \cdot \Phi_{mnk} + e^l \gamma^{nk} \cdot \Phi_{mnkl} + \dots \right]$$

$$\Rightarrow \underset{\text{total}}{d\xi_n} = \underset{\text{classic}}{dx_n} + \underset{\text{quantum}}{dy_n} = \underset{\text{classic}}{dx_n} + \underset{\text{quantum}}{dK\left(f_n\left(x\right)\right)}$$

$$\Rightarrow \underset{\text{total}}{\xi_n} = \underset{\text{classic}}{x_n} + \underset{\text{quantum}}{y_n} = \underset{\text{classic}}{x_n}$$

$$+ \overbrace{\left(p \cdot \Phi_m + p^n \cdot \Phi_{mn} + p^{nk} \cdot \Phi_{mnk} + p^{nkl} \cdot \Phi_{mnkl} + \dots \right)}^{\text{quantum}} \quad (194)$$

we would have an interesting way to introduce discrete masses M as matrix components via (193) and obtain

$$p \cdot \Phi_m + p^n \cdot \Phi_{mn} + p^{nk} \cdot \Phi_{mnk} + p^{nkl} \cdot \Phi_{mnkl} + \dots$$

$$= \frac{c \cdot \sqrt{2}}{\hbar} \cdot \left(M \cdot \Phi_m + M^n \cdot \Phi_{mn} + M^{nk} \cdot \Phi_{mnk} + M^{nkl} \cdot \Phi_{mnkl} + \dots \right). \quad (195)$$

Instead of stripping apart the masses, one could also apply the expansion to the Planck constant and obtain Planck vectors and tensors of various orders. This, however, should only be understood as general remark, not as a suggestion. The author clearly prefers the mass-expansion as a measure to try out first.

Summing Up the Generalized Recipe: The Forward Derivation

We intend to find the quantum part of a metric γ^{ij}, which is to say we search for the quantum field connected with this metric.

(1) We separate the coordinates of this metric $\xi_n = \xi_n^{\text{total}}$ into a classical and a quantum related part: $\xi_n^{\text{total}} = x_n^{\text{classic}} + y_n^{\text{quantum}} = x_n^{\text{classic}} + F\left(f_n\left(x\right)\right) \cdot f_n\left(x\right)^{\text{quantum}}$. Here F simply stands for a function of the function vector f_m and in many cases this is just a linear connector like: $K\left(f_n\left(x\right)\right) = F\left(f_n\left(x\right)\right) \cdot f_n\left(x\right) = p_n \cdot f_n\left(x\right) = \{p_0 \cdot f_0, \ p_1 \cdot f_1, \ p_2 \cdot f_2, \ p_3 \cdot f_3\}$. This gives the linear coordinate elements as: $d\xi_n^{\text{total}} = dx_n^{\text{classic}} + dy_n^{\text{quantum}} = dx_n^{\text{classic}} + dK\left(f_n\left(x\right)\right)^{\text{quantum}}$.

(2) We introduce these elements into the "clever-zero" equation of two total line elements as follows:

$$\Xi\left(ds_{\text{total}}^2\right) - \Xi\left(ds_{\text{total}}^2\right) = \Xi\left(\gamma^{nm} d\xi_n d\xi_m\right) - \Xi\left(\gamma_{nm} d\xi^n d\xi^m\right)$$

and substitute accordingly what we have for the covariant elements $d\xi_n$ plus set a generalized gradient expression according to (190) for their contravariant brothers which in simplest form would look like $d\xi^n = dx^n + \gamma^{ni}\frac{\partial f}{\partial x^i}$, $d\xi^m = dx^m + \gamma^{mi}\frac{\partial f}{\partial x^i}$ and/or we potentially extend to generalized Planck-tensors according to (194).

(3) Forming the roots, respectively de-factorizing according to (163) to (173) in the V_Ω form (also see the examples given in section "Examples for other 'Vectorial Dirac Roots'"), we result in a distinct number (depending on the number of independent metric components) of generalized Dirac-like equations for the functions f_m. These are all systems of partial differential equations, which, in the simplest cases, are given as:

$$\widehat{V}_\Omega^m f_m = \left\{ g^{mi}\frac{\partial}{\partial y^i} \pm i \cdot \sqrt{g^{mi}} \cdot e_i \cdot \hat{F} \right\}_\Omega f_m. \tag{196}$$

(4) Forming the scalar product of the Ω forms or V_Ω forms in the following form $\widehat{\Xi}\left(\frac{1}{\sqrt{g}}\left\{\widehat{V}_\Omega^n\right\}\sqrt{g}\right)f_m$ gives us partial differential equations of various orders (order higher than in the V_Ω forms), which in the cases of $\Xi[X] = X^2$ are resulting in Klein–Gordon-like equations with $\frac{1}{\sqrt{g}}\left\{\widehat{V}_\Omega^n\right\}\sqrt{g} \cdot \left\{\widehat{V}_\Omega^m\right\}f_m f_n$ of the form (with the definition $f_n \cdot f_m = e_n \cdot e_m h^2 = g_{nm}h^2 \equiv \Psi \cdot g_{nm}$, where we suppressed the bold f in the following):

$$0 = \left[-\hat{F}^2 + \square_{\text{Metric}}\right]\Psi + \frac{\Psi}{4} \cdot \frac{g^{nm}}{\sqrt{g}}\partial_m\sqrt{g} \cdot g^{mi}\partial_i g_{nm}. \tag{197}$$

(5) Thereby, the functional operator $\hat{F}$ can be extracted from the energy momentum tensor being defined in the symmetric Hilbert form: $T^{\alpha\beta} = \frac{2}{\sqrt{-g}}\frac{\delta\left(\sqrt{-g}\cdot L_M\right)}{\delta g_{\alpha\beta}}$ and the Euler–Lagrange equations for L_M, which should be the Lagrange density for matter.

Summing Up the Generalized Recipe: The Backward Derivation

This time we start with a given energy–momentum tensor and intend to find the quantum part of a metric γ^{ij}, which is to say we search for the metric connected with this tensor. Thereby we assume the quantum field of the matter being completely given due to the energy–momentum tensor.

(1) We set $\gamma^{ij} = g^{ij}$ and extract the functional operator $\hat{F}$ from the energy momentum tensor being defined in the symmetric Hilbert form: $T^{\alpha\beta} = \frac{2}{\sqrt{-g}}\frac{\delta\left(\sqrt{-g}\cdot L_M\right)}{\delta g_{\alpha\beta}}$. Thereby, the Euler–Lagrange equations for L_M, which is the Lagrange density for matter, will always take on the form $0 =$

$$\left[-\Xi\left(\hat{F}\right) + D_{\text{Metric}} \right] \Psi + \tfrac{\Psi}{4} \cdot g^{nm} D_{\text{Metric}} g_{nm} \text{ (usually it is just the classical:}$$
$$0 = \left[-\hat{F}^2 + \Box_{\text{Metric}} \right] \Psi + \tfrac{\Psi}{4} \cdot \tfrac{g^{nm}}{\sqrt{g}} \partial_m \sqrt{g} \cdot g^{mi} \partial_i g_{nm} \text{ with } \mathbf{f}_n \cdot \mathbf{f}_m \triangleq \mathbf{e}_n \cdot \mathbf{e}_m h^2 =$$
$g_{nm} h^2 \equiv \Psi \cdot g_{nm}$). Thus, it gives us the desired functional operator $\hat{F}$.

(2) The next step is to apply the new concept of extracting the V_Ω root in the Ω-form as described in the theory section by assuming

$$g_{nm} \widehat{\Xi}\left(\tfrac{1}{\sqrt{g}} \left\{ \hat{V}_\Omega \right\}^n \sqrt{g} \right) \Psi = g_{nm} \widehat{\Xi}\left(\tfrac{1}{\sqrt{g}} \left\{ \hat{V}_\Omega \right\}^n \sqrt{g} \right) g_{nm} h^2 = 0$$
$$= 0 = \left[-\Xi\left(\hat{F}\right) + D_{\text{Metric}} \right] \Psi + \tfrac{\Psi}{4} \cdot g^{nm} D_{\text{Metric}} g_{nm}$$

This results in an Ω number of Dirac-like equations in the V_Ω form as given in (163) to (173) (see also the examples given in section "Examples for other 'Vectorial Dirac Roots' "). In the simplest case we will find

$$\hat{V}_\Omega^m f_m = \left\{ g^{mi} \frac{\partial}{\partial y^i} \pm i \cdot \sqrt{g^{mi}} \cdot e_i \cdot \hat{F} \right\}_\Omega f_m = 0. \tag{198}$$

(3) After solving these equations and considering the coordinates $\xi_n = \overset{\text{total}}{\xi_n}$ of the unknown metric γ^{ij} to be separable as follows: $\overset{\text{total}}{\xi_n} = \overset{\text{classic}}{x_n} + \overset{\text{quantum}}{y_n} = \overset{\text{classic}}{x_n} + \underbrace{F\left(f_n\left(x\right)\right) \cdot f_n\left(x\right)}_{\equiv K\left(f_n(x)\right)}$, we have found the quantum

part of the deformable coordinates to be $\overset{\text{quantum}}{y_n} = \overset{\text{quantum}}{K\left(f_n\left(x\right)\right)}$, which—this way—has originally being defined by the matter part of the Einstein–Hilbert action.

Backward Example: The Higgs Field Revisited (Extended Consideration from [15])

Here we will repeat the evaluation as performed in [15] (also see "The Recipe" in this book), but this time, we will incorporate virtual parameters and thus, will end up in eigenvalue equations.

We choose the Lagrange density L_M as a function of a scalar parameter w like:

$$L_M = \gamma^{ij}, w_i, w_j - \mu^2 w^2 + \lambda^2 w^4. \tag{199}$$

The next step is to build the Euler–Lagrange equations, which gives us

$$\Box_{\text{Metric}} w + \mu^2 w - \lambda^2 w^3 \equiv \Box_{\text{Metric}} w - F^2 \left[w \right] \equiv \left[\Box_{\text{Metric}} - \hat{F}^2 \right] w = 0 \tag{200}$$

and where we recognize the Klein–Gordon equation for the setting $\lambda = 0$.

Comparison with Eq. (70) would give us $w = f_m$ but as w, as per definition, has to be a scalar function we have to assume that the operator $\left[\Box_{\text{Metric}} - \hat{F}^2 \right]$

only acts on one f_m component at a time. We assume that the contribution of the term $\frac{\Psi}{4} \cdot \frac{g^{nm}}{\sqrt{g}} \partial_m \sqrt{g} \cdot g^{mi} \partial_i g_{nm}$ can be ignored. It would only change the metric to be applied within the further evaluation. As we keep it general for now, no harm is done in not considering the Laplacian of the metric.

The corresponding generalized Dirac equations are already given in (75), where we have sucked all factors into the constants. The equations read

$$\widehat{V}_{\Omega}^m f_m = \left\{ i \cdot \sqrt{g^{mj}} \cdot e_j \cdot (\mu \pm i \cdot \lambda \cdot f_m) \pm g^{mi} \frac{\partial}{\partial x^i} \right\}_{\Omega} f_m = 0$$

$$\Rightarrow \widehat{V}_{\Omega}^0 w = \left\{ i \cdot \sqrt{g^{0j}} \cdot e_j \cdot (\mu \pm i \cdot \lambda \cdot w) \pm g^{0i} \frac{\partial}{\partial x^i} \right\}_{\Omega} w = 0, \quad (201)$$

where we have assumed that the parameter w is only connected with the metric components $m - 0$. Please note that our choice of the index $m = 0$ is completely arbitrary and has—for the time being—nothing to do with the usual setting $x_0 = t = $ time. Further assuming only diagonal metrics we obtain

$$\widehat{V}_{\Omega}^0 w\left(x^0\right) = \left\{ \begin{array}{l} i \cdot \sqrt{g^{00}} \cdot \left(\mu \pm i \cdot \lambda \cdot w\left(x^0\right)\right) \pm g^{00} \frac{\partial}{\partial x^0} \\ \pm E_1 \pm e^{\frac{i\pi}{2}} E_1 \pm \ldots \pm e^{\frac{(2j-1)i\pi}{2 \cdot j}} E_j \pm \ldots \end{array} \right\}_{\Omega} w\left(x^0\right) = 0,$$

$$(202)$$

which results in the following differential equations

$$\left\{ \begin{array}{l} i \cdot \sqrt{g^{00}} \cdot \left(\mu + i \cdot \lambda \cdot w\left(x^0\right)\right) + g^{00} \frac{\partial}{\partial x^0} \\ \pm E_1 \pm e^{\frac{i\pi}{2}} E_1 \pm \ldots \pm e^{\frac{(2j-1)i\pi}{2 \cdot j}} E_j \pm \ldots \\ i \cdot \sqrt{g^{00}} \cdot \left(\mu - i \cdot \lambda \cdot w\left(x^0\right)\right) + g^{00} \frac{\partial}{\partial x^0} \pm \ldots \\ i \cdot \sqrt{g^{00}} \cdot \left(\mu + i \cdot \lambda \cdot w\left(x^0\right)\right) - g^{00} \frac{\partial}{\partial x^0} \pm \ldots \\ i \cdot \sqrt{g^{00}} \cdot \left(\mu - i \cdot \lambda \cdot w\left(x^0\right)\right) - g^{00} \frac{\partial}{\partial x^0} \pm \ldots \end{array} \right\}_{\Omega} w\left(x^0\right) = 0 \quad (203)$$

giving the results ($C[k]$ just being constants)

$$w\left(x^0\right) = \left\{ \begin{array}{cc} \sqrt{e^{\frac{2\left(e^{\frac{ij\pi}{p}} E_p + i\mu\right)\left(x^0 + \sqrt{g^{00}}C[1]\right)}{\sqrt{g^{00}}}}} \left(e^{\frac{ij\pi}{p}} E_p - i\mu\right)^2 \pm \lambda \cdot e^{\frac{2e^{\frac{ij\pi}{p}} E_p\left(x^0 + \sqrt{g^{00}}C[1]\right)}{\sqrt{g^{00}}}} \left(e^{\frac{ij\pi}{p}} E_p - i\mu\right) \\ \hline e^{2i\mu\left(\frac{x^0}{\sqrt{g^{00}}} + C[1]\right)} - e^{\frac{2e^{\frac{ij\pi}{p}} E_p\left(x^0 + \sqrt{g^{00}}C[1]\right)}{\sqrt{g^{00}}}} \lambda^2 \\[3em] \pm \lambda \cdot e^{\frac{2e^{\frac{ij\pi}{p}} E_p\left(x^0 + \sqrt{g^{00}}C[1]\right)}{\sqrt{g^{00}}}} \left(e^{\frac{ij\pi}{p}} E_p - i\mu\right) - \sqrt{e^{\frac{2\left(e^{\frac{ij\pi}{p}} E_p + i\mu\right)\left(x^0 + \sqrt{g^{00}}C[1]\right)}{\sqrt{g^{00}}}}} \left(e^{\frac{ij\pi}{p}} E_p - i\mu\right)^2 \\ \hline e^{2i\mu\left(\frac{x^0}{\sqrt{g^{00}}} + C[1]\right)} - e^{\frac{2e^{\frac{ij\pi}{p}} E_p\left(x^0 + \sqrt{g^{00}}C[1]\right)}{\sqrt{g^{00}}}} \lambda^2 \\[3em] \frac{e^{\frac{ij\pi}{p}} E_p - i\mu}{e^{\frac{\left(e^{\frac{ij\pi}{p}} E_p - i\mu\right)\left(x^0 - \sqrt{g^{00}}C[1]\right)}{\sqrt{g^{00}}}} \mp \lambda} \quad , \quad \frac{i\mu - e^{\frac{ij\pi}{p}} E_p}{e^{\frac{\left(e^{\frac{ij\pi}{p}} E_p - i\mu\right)\left(x^0 - \sqrt{g^{00}}C[1]\right)}{\sqrt{g^{00}}}} \pm \lambda} \end{array} \right\}_{\Omega}$$

$$(204)$$

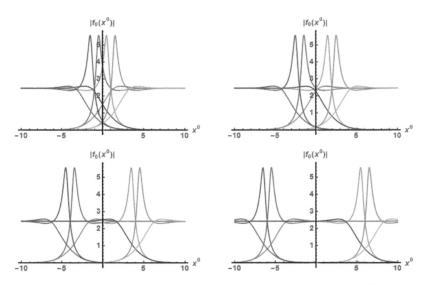

Figure 12 Various solutions as given in (204) with $\mu = 0.7$, $\lambda = 0.5$, $g^{00} = 1$, $P = 1$, $E_P = 1$ and $C[1] = 1$ (upper left), 2 (upper right), 4 (lower left) and 6 (lower right).

Here the parameter P is connected with the order of "virtuality" and j denotes the running index within such an order. For instance, in second order we have $P = 2$ and $j = 0$, 1, 2, 3 while for third and fourth order we find $P = 3$ and $j = 0$, 1, 2, 3, 4, 5 and $P = 4$ and $j = 0$, 1, 2, 3, 4, 5, 6, 7, respectively. In general we have $P =$ "number of order of virtuality" with the running index $j = 0$, 1, 2,..., $2*P - 1$. Please note that $P = 1$ also gives an eigenvalue equation but not a virtual parameter.

In order to generally illustrate some properties of the solutions above we will not bother about its connection to the Higgs field and just chose parameters such that certain aspects become easily visible. So, Figs. 12 to 14 show the various solutions for the "virtuality levels" 1 and 2, where we have also varied the integrational constant C[1]. We see that with increasing C[1] the maxima are moving away from the origin $w = 0$ in positive and negative direction. Please note that in the case $P = 1$ the parameter E_P is not truly virtual but only gives the real or the imaginary E_P solutions for $P = 2$ (c.f. Figs. 12 and 13 with Fig. 14).

It is most interesting to see that the integrational constant acts like a second coordinate (time part) in a wave function, which can also be seen directly from equation (204). Thus, interpreting $C[1]$ as time we would have

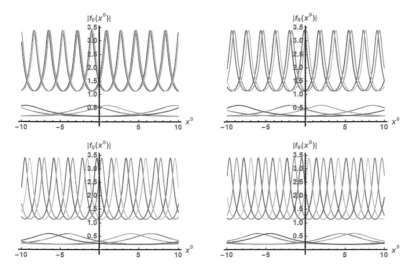

Figure 13 Various solutions as given in (204) with $\mu = 0.7$, $\lambda = 0.5$, $g^{00} = 1$, $P = 1$, $E_P = i = \sqrt{-1}$ and C[1] = 1 (upper left), 2 (upper right), 4 (lower left) and 6 (lower right).

an ordinary wave with speed $c = \sqrt{g^{00}}$. It will be discussed elsewhere whether this has something to do with our real time and the differences of matter and antimatter with respect to time.

Taking now $w = \xi_0{}^{\text{total}} = x_0{}^{\text{classic}} + y_0{}^{\text{quantum}} = x_0{}^{\text{classic}} + K\left(f_0\left(x = x_0 = g_{00}x^0\right)\right)^{\text{quantum}}$ we find the quantum distortion of the spatial coordinate x_0 caused by the Higgs field to be

$$\xi_0{}^{\text{total}} = x_0 + K\left(w\left(x^0\right)\right), \tag{205}$$

$$\xi_0{}^{\text{total}} = x_0 + \left\{\mu \cdot \left(w\left(x^0\right)\right) + i \cdot \lambda \cdot \left(w\left(x^0\right)\right)^2\right\}_\Omega. \tag{206}$$

Thus, we have obtained the spatial "quantum wiggle" causing the Higgs field for the coordinate x_0. The combination with the classical coordinate dimension is shown in Fig. 15 for the "virtuality" parameter $P = 3$;

With the solutions to the Higgs field known, we could now investigate what metrics would be necessary to create exactly the same set of solutions by sticking to the simple dependency $F(f_m) = p_m^* f_m$ as mainly used throughout our other previous papers. The background for this investigation has been discussed in [12] and is motivated by the fact that we could state that any change of reality should also mean a change of metric. Thus, all those

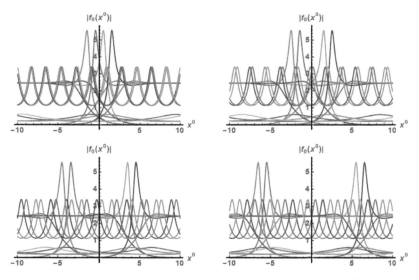

Figure 14 Various solutions as given in (204) with $\mu = 0.7$, $\lambda = 0.5$, $g^{00} = 1$, $P = 2$, $E_P = 1$ and $C[1] = 1$ (upper left), 2 (upper right), 4 (lower left) and 6 (lower right).

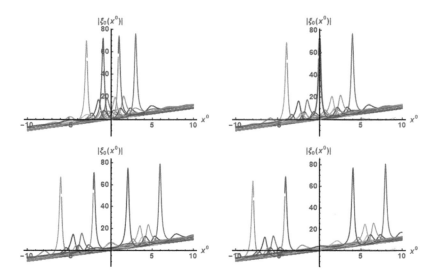

Figure 15 Various solutions as given in (204) in its consequence to the wiggles of the coordinate as given in (206) with $\mu = 0.7$, $\lambda = 0.5$, $g^{00} = 1$, $P = 3$, $E_P = 1$ and $C[1] = 1$ (upper left), 2 (upper right), 4 (lower left) and 6 (lower right).

various solutions given in (204) should also be considered various states of metrics instead of eigenstates to just one metric.

To investigate this we have to rewrite equation (202) as follows:

$$\hat{V}_{\Omega}^{0}w\left(x^{0}\right)=\left\{\begin{matrix} i\cdot\sqrt{g^{00}}\cdot\left(\mu\pm i\cdot\lambda\cdot w\left(x^{0}\right)\right)\pm g^{00}\frac{\partial}{\partial x^{0}}\\ \pm E_{1}\pm e^{\frac{i\pi}{2}}E_{1}\pm\ldots\pm e^{\frac{(2j-1)i\pi}{2\cdot j}}E_{j}\pm\ldots \end{matrix}\right\}_{\Omega}w\left(x^{0}\right)=0$$

$$\Rightarrow\tilde{V}_{\Omega}^{0}w\left(x^{0}\right)=\left\{\begin{matrix} \frac{i}{\sqrt{G^{00}}}\pm g^{00}\frac{\partial}{\partial x^{0}}\\ \pm\tilde{E}_{1}\pm e^{\frac{i\pi}{2}}\tilde{E}_{1}\pm\ldots\pm e^{\frac{(2j-1)i\pi}{2\cdot j}}\tilde{E}_{j}\pm\ldots \end{matrix}\right\}_{\Omega}w\left(x^{0}\right)=0$$

with $\dfrac{1}{\sqrt{G^{00}\left(x^{0}\right)}}=\dfrac{\left(\mu\pm i\cdot\lambda\cdot w\left(x^{0}\right)\right)}{\sqrt{g^{00}}}\Rightarrow G^{00}\left(x^{0}\right)$

$$=\frac{g^{00}}{\left(\mu\pm i\cdot\lambda\cdot w\left(x^{0}\right)\right)^{2}}.\tag{207}$$

Besides, it should be noted at this point that the new virtual eigenvalues $\tilde{E}_{i}$ reveal itself as the various levels of the background field, which will also be discussed in the next section in connection with the much simpler example of the harmonic quantum oscillator.

Evaluation of G^{00} and G_{00} for a variety of constellations is shown in the Figs. 16 to 20. The ground state with $\forall\tilde{E}_{i}=0$ is presented in Fig. 16. As

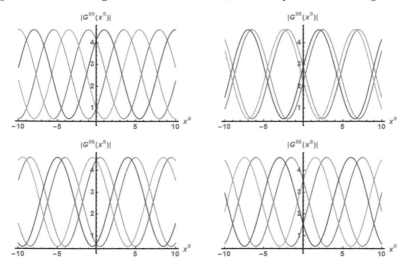

Figure 16 Various solutions for G^{00} as given in (207) with $\mu=0.7$, $\lambda=0.5$, $g^{00}=1$, $P=3$, $\forall\tilde{E}_{i}=0$ and C[1] = 1 (upper left), 2 (upper right), 4 (lower left) and 6 (lower right).

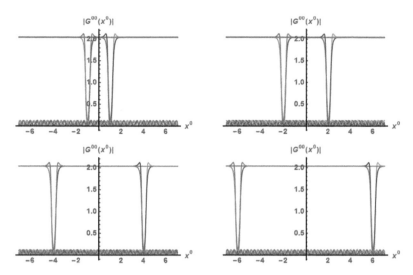

Figure 17 Various solutions for G^{00} as given in (207) with $\mu = 0.7$, $\lambda = 0.5$, $g^{00} = 1$, $P = 2$, $\tilde{E}_1 = 10$ and $C[1] = 1$ (upper left), 2 (upper right), 4 (lower left) and 6 (lower right).

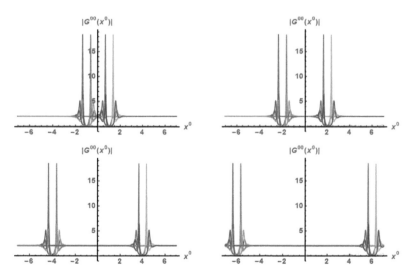

Figure 18 Various solutions for G^{00} as given in (207) with $\mu = 0.7$, $\lambda = 0.5$, $g^{00} = 1$, $P = 3$, $\tilde{E}_1 = 10$ and $C[1] = 1$ (upper left), 2 (upper right), 4 (lower left) and 6 (lower right).

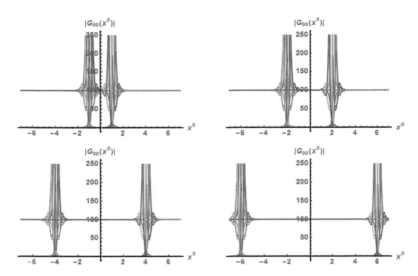

Figure 19 Various solutions for $G_{00} = 1/G^{00}$ as given in (207) with $\mu = 0.7$, $\lambda = 0.5$, $g^{00} = 1$, $P = 3$, $\tilde{E}_1 = 10$ and C[1] = 1 (upper left), 2 (upper right), 4 (lower left) and 6 (lower right).

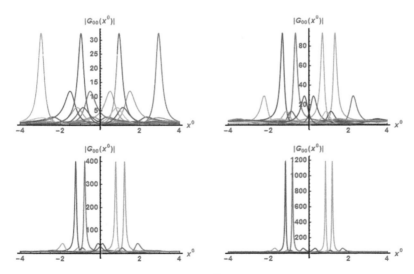

Figure 20 Various solutions for $G_{00} = 1/G^{00}$ as given in (207) with $\mu = 0.7$, $\lambda = 0.5$, $g^{00} = 1$, $P = 3$, C[1] = 1 and $\tilde{E}_1 = 1$ (upper left), $\tilde{E}_1 = 2$ (upper right), $\tilde{E}_1 = 3$ (lower left) and $\tilde{E}_1 = 4$ (lower right).

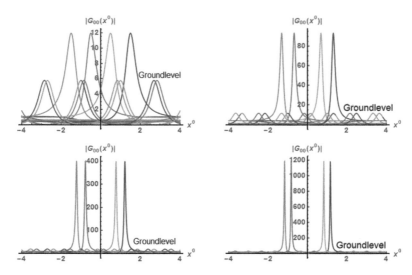

Figure 21 Various solutions for $G_{00} = 1/G^{00}$ as given in (207) with $\mu \approx 0.7$, $\lambda = 0.5$, $g^{00} = 1$, $P = 2$, $C[1] = 1$ and $\tilde{E}_1 = 1$ (upper left), $\tilde{E}_1 = 2$ (upper right), $\tilde{E}_1 = 3$ (lower left) and $\tilde{E}_1 = 4$ (lower right). This time we also marked the ground level.

one should expect one obtains stronger metric distortions for the covariant component G_{00} (Figs. 19 and 20) for bigger $\tilde{E}_i$ leading to weaker distortions for the inverse contravariant component G^{00} which was drawn in Figs. 17 and 18 for two levels of "virtuality." It can be seen (c.f. Fig. 21) that higher values of $\tilde{E}_i$ are leading to sharp distortions which maxima exceed the ground level in dependence of the magnitude of $\tilde{E}_i$. This effect is pretty sensitive to the parameter $\tilde{E}_i$. Thus, while for $P = 2$ and $\tilde{E}_i = 1$ the ground level almost reaches up to 50% of the distortional peaks, we have less than 1% for $\tilde{E}_i = 4$ with corresponding intermediate states for $\tilde{E}_i = 2$ and $\tilde{E}_i = 3$ (c.f. Fig. 21).

Forward Example: The Harmonic Oscillator and Eigenvalue Solutions for Simple Fields with $K(f_m) = F(f_m)^* f_m = p_m^* f_m$ Revisited (Extended Consideration from [10])

Here we repeat the evaluation as performed in [10], but this time, we will incorporate general virtual parameters.

For the simple functional $K(f_m) = p_m^* f_m$ we can make use of the evaluation of the previous section and set $\lambda = 0$. As this time we want to solve an eigenvalue problem, the subsequent generalized Dirac equations read

$$\hat{V}_\Omega w\left(x^0\right) = \left\{ \begin{matrix} i \cdot \sqrt{g^{00}} \cdot \mu + g^{00} \frac{\partial}{\partial x^0} \pm E_1 \pm e^{\frac{i\pi}{2}} E_1 \pm \dots \pm e^{\frac{(2j-1)i\pi}{2 \cdot j}} E_j \pm \dots \\ i \cdot \sqrt{g^{00}} \cdot \mu - g^{00} \frac{\partial}{\partial x^0} \pm E_1 \pm e^{\frac{i\pi}{2}} E_1 \pm \dots \pm e^{\frac{(2j-1)i\pi}{2 \cdot j}} E_j \pm \dots \end{matrix} \right\}_\Omega$$
$$\times w\left(x^0\right) = 0 \tag{208}$$

where simply out of curiosity we have assumed E to be a "virtual" parameter (compare with what has been said to equation (13)). The solutions are obtained by direct integration:

$$\overset{total}{\xi_0} = x_0 + F\left(w\left(x^0\right)\right) = x_0 + \left\{ p_0 \cdot \left(C[1]_\pm \cdot e^{\pm \frac{x^0\left(e^{\frac{ij\pi}{P}} E_P - i \cdot \mu\right)}{\sqrt{g^{00}}}} \right) \right\}_\Omega$$

$$= x_0 + \left\{ p_0 \cdot \left(e^{\pm C[2]_\pm \left(e^{\frac{ij\pi}{P}} E_P - i \cdot \mu\right)} \cdot e^{\pm \frac{x^0\left(e^{\frac{ij\pi}{P}} E_P - i \cdot \mu\right)}{\sqrt{g^{00}}}} \right) \right\}_\Omega$$

$$= x_0 + \left\{ p_0 \cdot e^{\pm \frac{\left(x^0 \pm \sqrt{g^{00}} \cdot C[2]_\pm\right)\left(e^{\frac{ij\pi}{P}} E_P - i \cdot \mu\right)}{\sqrt{g^{00}}}} \right\}_\Omega . \tag{209}$$

Again, we find a wave-like solution with the integration constant playing the role of one of the two wave coordinates in an assumed two dimensional space. It is been shown in [10] that in Kerr metrics this wave can be a revolution around a mass center.

Now we will see how we can obtain the Klein–Gordon quantum oscillator of our slightly more complex functional $F(w(x^0))$. Constructing the square of our V_Ω vectors as given in (84) we result in the Klein–Gordon equation for the harmonic quantum oscillator, which is also been obtained by equation (200) with $\lambda = 0$ and μ becoming a quadratic function of x^0, namely $\mu = \mu\left(x^0\right) = \mu_2\left(x^0\right)^2$. The subsequent equation now reads

$$\Box_{\text{Metric}} w + \mu^2 w - \lambda^2 w^3 \equiv \Box_{\text{Metric}} w - \frac{F^2[w]}{2} \equiv \left[\Box_{\text{Metric}} - \frac{\hat{F}^2}{2}\right] w = 0$$
$$\mu = \mu_2\left(x^0\right)^2; \quad \lambda = 0$$
$$\Rightarrow \left[\Box_{\text{Metric}} - \mu_2\left(x^0\right)^2\right] w\left(x^0\right) = 0 \tag{210}$$

and resembles the equation for the harmonic quantum oscillator.

Applying our usual V_Ω vectorial root extraction, we obtain a more general version of Eq. (84), which now reads

$$
\hat{V}_\Omega w\left(x^0\right) = \begin{cases} i \cdot \sqrt{g^{00}} \cdot \left(\mu_0 \pm \mu_1 \cdot \sqrt{x^0} \pm \mu_2 \cdot x^0 \pm \ldots\right) \pm g^{00} \frac{\partial}{\partial x^0} \\ \pm E_1 \pm e^{\frac{i\pi}{2}} E_1 \pm \ldots \pm e^{\frac{(2j-1)i\pi}{2j}} E_j \pm \ldots \end{cases}_\Omega
$$
$$
\times w\left(x^0\right) = 0. \tag{211}
$$

For the simple setting $\mu_0 = \mu_1 = 0$ and $\mu_2 \sim i \cdot \sqrt{g^{00}}$ we immediately obtain the ground state solution of the harmonic quantum oscillator.

It should be noted that, as performed in [12], one could also just assume that the metric "is taking on" all the coordinate dependencies, which is giving us back our simple $F(w)$ form but demands a more complex metric component g^{00}, being a function of the coordinate x^0. Then we simply assume eigenvalues different from the one in (87) and set

$$
\hat{V}_\Omega w\left(x^0\right) = \left\{ \frac{i}{\sqrt{g^{00}}} \pm \frac{\partial}{\partial x^0} \pm \tilde{E}_1 \pm e^{\frac{i\pi}{2}} \tilde{E}_1 \pm \ldots \pm e^{\frac{(2j-1)i\pi}{2j}} \tilde{E}_j \pm \ldots \right\}_\Omega w\left(x^0\right) = 0
$$
$$
\frac{1}{\sqrt{g^{00}}} = \left(\mu_0 \pm \mu_1 \cdot \sqrt{x^0} \pm \mu_2 \cdot x^0 \pm \ldots\right). \tag{212}
$$

Again we find the ground state solution for the harmonic quantum oscillator with the assumption of a metric of the form

$$
\frac{1}{\sqrt{g^{00}}} \sim i \cdot x^0 \Rightarrow \hat{V}_\Omega w\left(x^0\right)
$$
$$
= \left\{ -\mu \cdot x^0 \pm \frac{\partial}{\partial x^0} \pm \tilde{E}_1 \pm e^{\frac{i\pi}{2}} \tilde{E}_1 \pm \ldots \pm e^{\frac{(2j-1)i\pi}{2j}} \tilde{E}_j \pm \ldots \right\}_\Omega w\left(x^0\right) = 0
$$
$$
\tag{213}
$$

where μ has to be positive definite and the plus sign before the derivative will lead to solutions not converging for infinite x^0. Thus, the remaining equation for the ground state of the harmonic oscillator will be found for a metric $g^{00} \sim -\left(x^0\right)^{-2}$; $g_{00} \sim -\left(x^0\right)^2$ with the subsequent equation:

$$
\Rightarrow \hat{V}_\Omega w\left(x^0\right) = \left\{ -\mu \cdot x^0 - \frac{\partial}{\partial x^0} \pm \tilde{E}_1 \pm e^{\frac{i\pi}{2}} \tilde{E}_1 \pm \ldots \pm e^{\frac{(2j-1)i\pi}{2j}} \tilde{E}_j \pm \ldots \right\}_\Omega w\left(x^0\right) = 0. \tag{214}
$$

It will be derived elsewhere [16] how the metric can be obtained for all classical quantum states of the harmonic quantum oscillator (see also the previous chapter about the harmonic oscillator in 1D). It will also be shown that the classical model must be incomplete. Here we only give the result (with $H_n(x)$ denoting the Hermite polynomial of order n and argument x):

$$\Rightarrow \hat{V}_{n\Omega} w_n\left(x^0\right) = \left\{ \begin{array}{c} \mu \cdot \left(\dfrac{2 \cdot n \cdot H_{n-1}\left(x^0\right)}{H_n\left(x^0\right)} - x^0\right) - \dfrac{\partial}{\partial x^0} \\[2mm] \pm \tilde{E}_1 \pm e^{\frac{i\pi}{2}} \tilde{E}_1 \pm \ldots \pm e^{\frac{(2j-1)i\pi}{2 \cdot j}} \tilde{E}_j \pm \ldots \end{array} \right\}_\Omega w_n\left(x^0\right) = 0.$$

(215)

Important hint: One should not mix up the classical energy eigenstates of energy E_{cn} of the classical quantum mechanical solution given as $E_{cn} = \hbar\omega\left(\frac{1}{2} + n\right)$ with our virtual eigenvalues and eigensolutions resulting from non-zero $\tilde{E}_i$! These $\tilde{E}_i$ namely, have to be understood as true virtual parameters and thus, they are the equivalent for the so-called "second quantization" resulting from the perturbations caused by the background quantum field. Here, however, this background quantum field just reveals itself as the additional jitter of the space-time adding up on the metric's curvature.

Conclusions to "Generalization of the Recipe"

It was shown how any smooth space can be quantized via its line element in a very general manner. Along the way the classical Planck constant, which stands for the limit to the "quantum world," can be generalized to a Planck tensor or even a tensorial function.

The new method was applied to the Higgs field and interesting solutions are found for eigenvalue formulations, which might be interpreted as particles masses.

It was also found how the natural virtual parameters residing in the new quantization method stand as equivalent for the so-called second quantization and provide the background perturbation.

Chapter 7

About Fermat's Last Theorem

Introduction

At some point within the years between 1637 and 1643 Pierre de Fermat scribbled in the margin of his copy of the ancient Greek text *Arithmetica* by Diophantus that he has found a nice little proof to the fact that the equation $x^n + y^n = z^n$ does not have integer solutions for all n, x, y, z for $n > 2$. This theorem was not proven until 1994. The proof was found by Andrew Wiles and Richard Taylor.

Unfortunately, the proof found by Wiles and Taylor was by far too lengthy (almost 100 pages without attachments) and too complicated to have ever been the one Fermat might have meant in his statement. In addition, the techniques Wiles and Tayler had used were all results of modern mathematics and thus, Fermat could never had have access to such knowledge.

So, the community has asked quite often, did Fermat lie about his nice little proof or is there an alternative and much simpler way, nobody has seen so far?

The original text of Fermat's statement, written in Latin, reads (http://mathworld.wolfram.com/FermatsLastTheorem.html)

> Cubum autem in duos cubos, aut quadrato-quadratum in duos quadrato-quadratos, et generaliter nullam in infinitum ultra quadratum potestatem

The Theory of Everything: Quantum and Relativity is Everywhere – A Fermat Universe
Norbert Schwarzer
Copyright © 2020 Jenny Stanford Publishing Pte. Ltd.
ISBN 978-981-4774-47-5 (Hardcover), 978-1-315-09975-0 (eBook)
www.jennystanford.com

in duos eiusdem nominis fas est dividere cuius rei demonstrationem mirabilem sane detexi. Hanc marginis exiguitas non caperet.

The translation can also be found at mathworld:

It is impossible for a cube to be the sum of two cubes, a fourth power to be the sum of two fourth powers, or in general for any number that is a power greater than the second to be the sum of two like powers. I have discovered a truly marvelous demonstration of this proposition that this margin is too narrow to contain.

Motivation

With the assumption of time and space not being truly continuous, one has to realize that there must be a connection between Fermat's last theorem and quantum theory. This arises because at a certain scale, nature has to work with whole things/entities, which is to say it is forced to "calculate" with integers simply because there are neither fractals nor reals at hand anymore as neither time nor space are continuous and the smallest things cannot be divided any further. To formulate it differently: As quantum theory is based on the assumption of the existence of "principally smallest things" then these things must be whole and have to be counted in integers. Thus, any mathematical operation forcing nature to "apply" reals or fractals, must be forbidden at small scales.

On the other hand, it is clear that any general quantization of any smooth space must include the quantization of this space's line element (c.f. section "Further Considerations"). This connects quantum theory with Fermat's last theorem, because the line element has to be evaluated out of the space's metric, directly leading to squared expressions. Powers of higher orders are thinkable and can be constructed, even quantized (e.g. [9]), but they are not leading to the physics we know.

Why is That?

Perhaps because any other mathematical operation with powers bigger than 2 does not allow for integer solutions? As we have seen in our derivations of the new quantization methods as outlined in the Theory section, we very often are facing problems of the kind $x^n + y^n = z^n$ (c.f. previous sections and examples), with x and y standing for functions and z for eigenvalues.

The fact that these are functions should not change the condition for the need to guaranty the outcome of integer solutions for equations of this kind, because the functions, too, will finally have to give certain results in form of numbers.

If so, than there must be a much simpler way to proof Fermat's last theorem on the basis of quantum theory techniques than the one found by Wiles and Taylor in 1994. Indeed, when using the method to extract roots via the V_Ω vector as introduced in the theory section of this book, things are looking quite promising. However, as there have been many others who had tried to proof Fermat's last theorem over more than 350 years, the author still rather sees his attempt as a mere suggestion rather than a rigorous mathematical proof.

Fermat's Own Proof?

Following the method derived in the theory section and its generalization in section "The Generalized 'Vectorial Dirac Root,'" we introduce a vector V_Ω with components of all m possible tuples of combinations of phase terms with integer parameters j_k of the following form (c.f. Eq. (163)):

$$V_\Omega = \Upsilon \left\{ a_1 + a_2 \cdot e^{i \cdot j_1 \cdot \Delta\varphi_1} + \ldots + a_{k+1} \cdot e^{i \cdot j_k \cdot \Delta\varphi_k} \right\}$$

$$\equiv \left(\frac{1}{m = k^n}\right)^{\frac{1}{n}} \left\{ \begin{array}{l} a_1 + a_2 + \ldots + a_{k+1} \\ a_1 + a_2 \cdot e^{i \cdot \Delta\varphi_1} + \ldots + a_{k+1} \\ \vdots \\ a_1 + a_2 \cdot e^{i \cdot j_1 \cdot \Delta\varphi_1} + \ldots + a_{k+1} \cdot e^{i \cdot j_k \cdot \Delta\varphi_k} \\ \vdots \\ a_1 + a_2 \cdot e^{i \cdot (n-1) \cdot \Delta\varphi_1} + \ldots + a_{k+1} \cdot e^{i \cdot (n-1) \cdot \Delta\varphi_k} \end{array} \right\} \quad (216)$$

where we have to set $\Delta\varphi_s = \frac{2\pi}{n}$. We find that for integer $n > 1$ we can prove that

$$[V_\Omega]^n = \frac{1}{k} \sum_{j_1,\ldots, j_k=0}^{n-1} \left(a_1 + a_2 \cdot e^{i \cdot j_1 \cdot \Delta\varphi_1} + \ldots + a_{k+1} \cdot e^{i \cdot j_k \cdot \Delta\varphi_k}\right)^n$$

$$= (a_1)^n + \ldots + (a_{k+1})^n ; \quad \Delta\varphi_s = \frac{2\pi}{n}. \quad (217)$$

Fermat might not have had complex numbers, but perhaps he just used a similar expression with two orthogonal components. In his case the vector decomposition as given above would read

$$[V_\Omega]^n = \frac{1}{2} \sum_{j_1, j_2=0}^{n-1} \left(x + y \cdot e^{i \cdot j_1 \cdot \frac{2\pi}{n}} + (-1)^{\frac{1}{n}} z \cdot e^{i \cdot j_2 \cdot \frac{2\pi}{n}} \right)^n = x^n + y^n - z^n = 0.$$

(218)

Integer solutions of (218) for x, y, z and integer n (x, y, z, $n \ \varepsilon \ Z$) now require that all the vector components of V_Ω also give integers, which leads to 2^n equations of the form

$$x + y \cdot e^{i \cdot j_1 \cdot \frac{2\pi}{n}} + (-1)^{\frac{1}{n}} z \cdot e^{i \cdot j_2 \cdot \frac{2\pi}{n}} \Big|_{\forall j_1, j_2 = \{0,1,2,\dots,n-1\}} = \tilde{Q}_{j_1, j_2} + (-1)^{\frac{1}{n}} \tilde{P}_{j_1, j_2}$$

$$\Rightarrow y \cdot e^{i \cdot j_1 \cdot \frac{2\pi}{n}} + (-1)^{\frac{1}{n}} z \cdot e^{i \cdot j_2 \cdot \frac{2\pi}{n}} \Big|_{\forall j_1, j_2 = \{0,1,2,\dots,n-1\}} = Q_{j_1} + (-1)^{\frac{1}{n}} P_{j_2}$$

$$\Rightarrow y \cdot e^{i \cdot j_1 \cdot \frac{2\pi}{n}} \Big|_{\forall j_1, j_2 = \{0,1,2,\dots,n-1\}} = Q_{j_1} \quad \text{and} \quad z \cdot e^{i \cdot j_2 \cdot \frac{2\pi}{n}} \Big|_{\forall j_1, j_2 = \{0,1,2,\dots,n-1\}}$$

$$- P_{j_2} \tilde{Q}_{j_1, j_2} , \tilde{P}_{j_1, j_2}, Q_{j_1, j_2}, P_{j_1, j_2} \in Z.$$

(219)

One easily sees now, that this is only possible for $n = 2$, because natural (non-complex) integer numbers for x, y and z are only obtained from the phase factors $e^{i \cdot j_1 \cdot \frac{2\pi}{n}}$, $e^{i \cdot j_2 \cdot \frac{2\pi}{n}}$ for $n = 2$, while this is impossible for all $n > 2$.

Yes indeed, this might have been too much for Fermat's ominous margin, especially as he probably had to use somewhat longer expressions for the complex exponential phase factors, but in the opinion of this author, it comes close enough to an acceptable length for Fermat's own original idea. As said before, we leave it to the community to find the flaw.

Chapter 8

Dirac Quantization of the Kerr Metric

The Generalized Metric Dirac Operator for a Kerr Object "at Rest"

Applying the technique as proposed in the theory section, the Kerr metric is given in the form [18, 19]:

$$g^{ij} = \begin{pmatrix} g^{00} & 0 & 0 & g^{30} \\ 0 & g^{11} & 0 & 0 \\ 0 & 0 & g^{22} & 0 \\ g^{30} & 0 & 0 & g^{33} \end{pmatrix} \xrightarrow{\text{Kerr}}$$

$$= \begin{pmatrix} g^{00} & 0 & 0 & g^{30} \\ 0 & -\frac{\Delta}{r_n^2 + \alpha^2 \cos[\theta]^2} & 0 & 0 \\ 0 & 0 & -\frac{1}{r_n^2 + \alpha^2 \cos[\theta]^2} & 0 \\ g^{30} & 0 & 0 & \frac{-1}{\Delta \cdot \sin[\theta]^2} \left(1 - \frac{r_s \cdot r_n}{\rho^2}\right) \end{pmatrix}$$

with $\quad \Delta = r_n^2 - r_s \cdot r_n + \alpha^2; \quad r_s = \dfrac{2MG}{c^2}; \quad \alpha = \dfrac{J}{M \cdot c};$

$$\rho = \sqrt{r_n^2 + \alpha^2 \cos[\theta]^2} \tag{220}$$

The Theory of Everything: Quantum and Relativity is Everywhere – A Fermat Universe
Norbert Schwarzer
Copyright © 2020 Jenny Stanford Publishing Pte. Ltd.
ISBN 978-981-4774-47-5 (Hardcover), 978-1-315-09975-0 (eBook)
www.jennystanford.com

and

$$g^{00}(r_n) = \frac{\frac{J^2}{c^2 M^2} + r_n^2 + \frac{2GJ^2 M r_n \sin[\theta]^2}{c^4 M^2 r_n^2 + c^2 J^2 \cos[\theta]^2}}{\frac{J^2}{M^2} + r_n(-2GM + c^2 r_n)}; \quad r_n = |\vec{r} - \vec{r}_n|$$

$$g^{30}(r_n) = \frac{4GJ M^4 r_n}{(J^2 + M^2 r_n(-2GM + c^2 r_n))(c^2 M^2 r_n^2 + J^2 \cos[\theta]^2)} \quad (221)$$

leads us to a set of partial differential equations for the component $f = f_{(0)}$ as follows:

$$V_\Omega = \begin{cases} \alpha + \beta + (\gamma + \delta) + a + \overbrace{b + c}^{=\,0} + d, \alpha + \beta + i(\gamma - \delta) + a + b + c + d, \\ \alpha + \beta - i(\gamma - \delta) + a + b + c + d, \alpha + \beta - (\gamma + \delta) + a + b + c + d, \\ \alpha - \beta + (\gamma + \delta) + a + b + c + d, \alpha - \beta + i(\gamma - \delta) + a + b + c + d, \\ \alpha - \beta - i(\gamma - \delta) + a + b + c + d, \alpha - \beta - (\gamma + \delta) + a + b + c + d, \\ \alpha + \beta + (\gamma + \delta) - (a + b + c + d), \alpha + \beta + i(\gamma - \delta) - (a + b + c + d), \\ \alpha + \beta - i(\gamma - \delta) - (a + b + c + d), \alpha + \beta - (\gamma + \delta) - (a + b + c + d), \\ \alpha - \beta + (\gamma + \delta) - (a + b + c + d), \alpha - \beta + i(\gamma - \delta) - (a + b + c + d), \\ \alpha - \beta - i(\gamma - \delta) - (a + b + c + d), \alpha - \beta - (\gamma + \delta) - (a + b + c + d) \end{cases} = 0.$$

$$(222)$$

Here we have set

$$\alpha = i \cdot \sqrt{g^{00}} \cdot \mathbf{f}_0; \ \beta = i \cdot \sqrt{g^{33}} \cdot \mathbf{f}_3; \ \gamma = i \cdot \sqrt{g^{30}} \cdot \mathbf{f}_0; \ \delta = i \cdot \sqrt{g^{30}} \cdot \mathbf{f}_3;$$

$$a = g^{00} \cdot \frac{\partial \mathbf{f}_0}{\partial x^0}; \ b = g^{33} \cdot \frac{\partial \mathbf{f}_3}{\partial x^3}; \ c = g^{30} \cdot \frac{\partial \mathbf{f}_3}{\partial x^0}; \ d = g^{30} \cdot \frac{\partial \mathbf{f}_0}{\partial x^3} \quad (223)$$

and have—for brevity—ignored the f_3 components with respect to the derivatives. A full solution will be given further below. The solutions are easily found for a particle "at rest," because we assume the functions f_i only to be dependent on time and the angle of rotation φ. Fortunately, the metric components of the Kerr metric do not contain these coordinates, which leaves the equation quite simple. We obtain functions of the form

$$f_{(0)}(x^0, x^3) = f(t, \phi); \quad \pm I = \{+, -, i, -i\};$$

$$f \to \text{Function}\left[e^{\pm \frac{\left(i\sqrt{g^{00}} \pm I \cdot \sqrt{g^{30}}\right) \cdot t}{g^{00}}} \cdot C \cdot \left[\frac{g^{00} \cdot \phi - g^{30} \cdot t}{g^{00}} \right] \right], \quad (224)$$

where we not only recognize the exponential part as the classical Dirac term, but also discover an additional characteristic wave function in t and φ of arbitrary form with the argument $\left[\frac{g^{00} \cdot \phi - g^{30} \cdot t}{g^{00}} \right]$. Thus, particular solutions could simply be sin- or cos-waves (with arbitrary ω)

$$f(t, \phi) = e^{\pm \frac{\left(i\sqrt{g^{00}} \pm I \cdot \sqrt{g^{30}}\right) \cdot t}{g^{00}}} \cdot C \cdot \begin{cases} \sin\left[\omega \cdot \left(\phi - \frac{g^{30} \cdot t}{g^{00}}\right)\right] \\ \cos\left[\omega \cdot \left(\phi - \frac{g^{30} \cdot t}{g^{00}}\right)\right] \end{cases}. \quad (225)$$

Interpreting, as Dirac has done, the solutions as particles, one might finally understand the spin as a revolving wave permanently orbiting a center. This center would then also be the center of the particle. With the interpretation of the revolving wave as spin we also have to allow for two spin directions $+\,|\,g^{30}|$ and $-\,|\,g^{30}\,|$, which simply means $+\,|\,J\,|$ and $-\,|\,J\,|$ for our Kerr object (c.f. Eqs. (237) and (221)).

In order to satisfy the condition that Lepton as spin-1/2 particles reproduce their state after two rotations, we simply demand the exponent of the e function to be half the wave argument for a fixed φ. We obtain the equation

$$\frac{\left(\sqrt{g^{00}}\pm I\sqrt{g^{30}}\right)}{g^{00}}=\frac{\omega\cdot g^{30}}{2\cdot g^{00}}$$

$$\Rightarrow g^{30}=\frac{2}{\omega}\left((\pm I)^2\pm\sqrt{(\pm I)^4+(\pm I)^2\,2\omega\cdot\sqrt{g^{00}}}+\omega\cdot\sqrt{g^{00}}\right). \quad (226)$$

Interestingly, this gives a $\pm\,|\,g^{30}|$ situation with equal but opposite spin states only in the case of $(\pm I)^2+\omega\cdot\sqrt{g^{00}}=0\Rightarrow\omega=-\frac{(\pm I)^2}{\sqrt{g^{00}}}$ resulting in the "spin shear elements" $g^{30}=\pm\frac{2\cdot i\cdot g^{00}}{(\pm I)^2}=\pm 2\cdot i\cdot g^{00}$. As the metric components g^{30} and g^{00} are complexly dependent on the two coordinates r and θ it is not clear to the author how such a simple dependency could be satisfied respectively "where" it must be fulfilled. The only explanation, if the Kerr particle does present Leptons, would be that the classical spin description and handling is just an approximation of the true reality.

With respect to function f_3 we obtain the following solution:

$$f_{(3)}\left(x^0,x^3\right)=F\,(t,\phi);\quad\pm I=\{+,-,i,-i\};$$

$$F\to\text{Function}\left[e^{\pm\frac{\left(i\sqrt{g^{33}}\pm I\cdot\sqrt{g^{30}}\right)\cdot t}{g^{30}}}\cdot C\cdot\left[\phi-\frac{g^{33}\cdot t}{g^{30}}\right]\right]. \quad (227)$$

By the means of virtual parameters as introduced e.g. in [10], the equations (238) can easily be made an eigenvalue equations with respect to certain eigenvalues E. Their solutions would be

$$f_{(0)}\left(x^0,x^3\right)=f\,(t,\phi);\quad\pm I=\{+,-,i,-i\};$$

$$f\to\text{Function}\left[e^{\pm\frac{\left(i\sqrt{g^{00}}\pm I\cdot\sqrt{g^{30}}\pm I\cdot E_0\right)\cdot t}{g^{00}}}\cdot C_0\cdot\left[\phi-\frac{g^{30}\cdot t}{g^{00}}\right]\right], \quad (228)$$

and

$$f_{(3)}\left(x^0, x^3\right) = F\left(t, \phi\right); \quad \pm I = \{+, -, i, -i\};$$

$$F \to \text{Function}\left[e^{\pm \frac{\left(i\sqrt{g^{33}}\pm I \cdot \sqrt{g^{30}} \pm I \cdot E_3\right) \cdot t}{g^{30}}} \cdot C_3 \cdot \left[\phi - \frac{g^{33} \cdot t}{g^{30}}\right]\right]. \quad (229)$$

The Kerr object at rest would then be characterized by the vector

$$\mathbf{f}\left(x^0, x^3\right) = \left\{\begin{array}{c} f_{(0)}\left(x^0, x^3\right) \\ f_{(3)}\left(x^0, x^3\right) \end{array}\right\}$$

$$= \left\{\begin{array}{c} \text{Function}\left[e^{\frac{\left(\left\{\begin{smallmatrix} - & - & - & - \\ - & - & - & - \\ + & + & + & + \\ + & + & + & + \end{smallmatrix}\right\} i\sqrt{g^{00}} + \left\{\begin{smallmatrix} -i & - & + & i \\ -i & + & - & i \\ i & - & + & -i \\ +i & - & + & -i \end{smallmatrix}\right\} \cdot \sqrt{g^{30}} \pm I \cdot E_0\right) \cdot t}{g^{00}}} \cdot C_0 \cdot \left[\phi - \frac{g^{30} \cdot t}{g^{00}}\right]\right] \\ \\ \text{Function}\left[e^{\frac{\left(\left\{\begin{smallmatrix} - & - & - & - \\ + & + & + & + \\ + & + & + & + \\ - & - & - & - \end{smallmatrix}\right\} i\sqrt{g^{33}} + \left\{\begin{smallmatrix} -i & - & + & i \\ -i & - & + & i \\ i & + & - & -i \\ +i & + & - & -i \end{smallmatrix}\right\} \cdot \sqrt{g^{30}} \pm I \cdot E_3\right) \cdot t}{g^{30}}} \cdot C_3 \cdot \left[\phi - \frac{g^{33} \cdot t}{g^{30}}\right]\right] \end{array}\right\}.$$

$$(230)$$

Ignoring the exponential term one might also demand the 4 Pi character of the leptons by setting either $\frac{g^{30} \cdot t}{g^{00}} = \frac{g^{33} \cdot t}{2g^{30}}$ or $\frac{g^{30} \cdot t}{2g^{00}} = \frac{g^{33} \cdot t}{g^{30}}$. The results would be $\frac{\left(g^{30}\right)^2 \cdot 2}{g^{00}} = g^{33}$ or $\frac{\left(g^{30}\right)^2}{2g^{00}} = g^{33}$.

Further Results and Trials

It is pretty illustrative to consider the term $\left[\phi - \frac{g^{30} \cdot t}{g^{00}}\right]$ with respect to the characteristic time T necessary to finish a complete revolution at the Schwarzschild radius r_s of an electron (myon or tauon) around its equator. Assuming the equator velocity of the spinning Kerr–Newman object to be the speed of light in vacuum $v = c$ and setting

$$
\left.\frac{g^{30}}{g^{00}}\right|_{r=r_s, \theta=\pi/2} \cdot T = 2 \cdot \pi = \left.\frac{g^{30}}{g^{00}}\right|_{r=r_s, \theta=\pi/2} \cdot \frac{s}{c} \tag{231}
$$

we find

$$
\alpha = \frac{s}{\pi} \quad \Rightarrow \quad 1 = \left.\frac{g^{30}}{g^{00}}\right|_{r=r_s, \theta=\pi/2} \cdot \frac{\alpha}{2 \cdot c}
$$

$$
1 = \left[\frac{2c\left(r_s^2 - r_Q^2\right)\alpha}{r_s^4 - r_Q^2\alpha^2 + r_s^2\alpha^2 + r_s^2\alpha^2}\right] \cdot \frac{\alpha}{2 \cdot c} = \left[\frac{2\left(r_s^2 - r_Q^2\right)\alpha}{r_s^4 - r_Q^2\alpha^2 + r_s^2\alpha^2 + r_s^2\alpha^2}\right] \cdot \frac{\alpha}{2}.
$$
$$\tag{232}$$

Here we had to take into account that r_s and α are connected, namely via $\alpha = \frac{2 \cdot J \cdot G}{c^3 \cdot r_s}$. Thus, the characteristic length s our rotating wave has propagated with the speed of light along a certain equator-position of the Kerr–Newman object would be completely disconnected from the position of $r = r_s$ we have chosen out of curiosity. In order to find a reasonable connection we should rather set

$$
\left.\frac{g^{30}}{g^{00}}\right|_{r=n\cdot r_s, \theta=\pi/2} \cdot T = 2 \cdot \pi = \left.\frac{g^{30}}{g^{00}}\right|_{r=n\cdot r_s, \theta=\pi/2} \cdot \frac{n \cdot r_s}{c}, \tag{233}
$$

which in fact immediately results in much more reasonable values for the equator position with the propagation-speed condition $v = c$. The equation to solve would now be

$$
\left.\frac{g^{30}}{g^{00}}\right|_{r=n\cdot r_s, \theta=\pi/2} \cdot T = 2 \cdot \pi = \left[\frac{2c\left(n \cdot r_s^2 - r_Q^2\right)\alpha}{n^4 r_s^4 - r_Q^2\alpha^2 + n^2 r_s^2\alpha^2 + n \cdot r_s^2\alpha^2}\right] \cdot \frac{n \cdot r_s}{c}.
$$
$$\tag{234}$$

From the six roots we have 4 complex ones which sum up to two purely real positive and negative Schwarzschild radii of the electron for $n = 4{,}9406 * 10^{44}$. This gives $n*r_s = 6{,}684 * 10^{-13}$ which is almost exactly 2 times the specific total angular momentum parameter α. The two remaining roots are completely imaginary and of opposite sign. Their magnitude is suspiciously close to the expected values for the electron neutrino but we find that the roots disappear the moment we set the charge parameter $r_Q = 0$ as we should when considering a neutrino.

Things are changing, however, if we introduce a cosmological constant and apply the Kerr–Newman–de Sitter solution [28–30] instead of the pure

Kerr–Newman one. The equation to solve now becomes

$$\frac{g^{30}}{g^{00}}\bigg|_{r=n\cdot r_s,\theta=\pi/2} \cdot T = 2 \cdot \pi$$

$$= \left[\frac{4c^3 G J\, r_s \left(4G^2 J^2 n^2 \Lambda + c^6 \left(-3r_Q^2 + 3nr_s^2 + n^4 r_s^4 \Lambda\right)\right)}{3c^{12} n^4 r_s^6 + 16 G^4 J^4 n^2 \Lambda + 4c^6 G^2 J^2 \left(-3r_Q^2 + nr_s^2 \left(3 + 3n + n^3 r_s^2 \Lambda\right)\right)} \right]$$

$$\times \frac{n \cdot r_s}{c}. \tag{235}$$

Now the two additional roots are becoming real (still positive and negative, which stands for matter and antimatter), they do not disappear for zero charge solutions and we still obtain rather intriguing values for all— supposed to be—lepton particles (c.f. table below). It should be noted that the results are very sensitive with respect to the input parameters, especially the angular momentum J. In order to avoid long-winded discussions on this topic the author has performed the evaluation for both the total angular momentum and the component angular momentum (see Table 1).

The cosmological constant we have applied was $\Lambda = 10^{-52}\ \mathrm{m}^{-2}$ (non-zero and positive).

Table 1   Results for "neutrino-masses" (?) with n evaluated such that the right masses for the charged leptons would come out of equation (235). Then the *n* found "where put into the equation again" and varied over the region 0.1*n to 10*n of the previous result in order to see the evolution of the other roots. While four roots where only various forms of the charged leptons again we also found two real roots with much lower masses respectively Schwarzschild radii. We interpret these other masses as the masses of the neutrinos and give the range in which these apparent neutrino masses have varied when varying the input parameter *n* as said above (namely from 0.1*n to 10*n)

Particle pair	Electron/ electron-neutrino	Myon/ myon-neutrino	Tauon/ tauon-neutrino
Mass for charged particle in MeV	0,511	105,66	1777
Interpreted mass variation for neutrino in eV with total angular momentum $J = \hbar\frac{\sqrt{1\cdot(1+1)}}{2} = \hbar\frac{\sqrt{3}}{2}$	0,0025 – 0,0215	0,4 – 3,2	7 – 46
Interpreted mass for neutrino in eV with component angular momentum $J = \hbar\frac{1}{2}$	0,0015 – 0,0113	0,45 – 3,25	6 – 57

It should be pointed out explicitly that the "neutrino masses" respectively their ranges derived here are based on the assumption of a clear similarity between the revolving wave propagation for the charged and the neutral leptons. For both particle types the radius to reach a certain speed $v = c$ for the revolving equator (a fictive position only to set the boundary $v = c$) was assumed to be in the same range in measures of the Schwarzschild radius of the objects in question. This does not necessarily mean that the author assumes the equator of the object at this very position or to actually rotate at this speed (propagate more like, because we are talking about a characteristic wave-solution). No, it is only a fictive distance from the particle's center to define a certain boundary condition, namely the one given in (233). The question of the true size and spatial distribution or appearance of the leptons still needs to be answered. We will do this in the next section.

The Spatial Appearance of the Leptons

In order to obtain an illustrative understanding about the leptons and their distribution in space we resort to equation (52) which is connecting the metric coordinates with the quantum mechanical fluctuations. As a first application, we consider only the metric component for the radius g^{11} and again apply the Kerr–Newman metric. Even though the equation is relatively simple and of first order

$$\left\{ \begin{array}{l} i \cdot \mu + \sqrt{g^{11}\left(x^1, x^2\right)} \frac{\partial}{\partial x^1} \\ i \cdot \mu - \sqrt{g^{11}\left(x^1, x^2\right)} \frac{\partial}{\partial x^1} \end{array} \right\}_{\Omega} f\left(x^1, x^2\right) = \pm I \cdot E \cdot f\left(x^1, x^2\right), \quad (236)$$

the subsequent solutions are not, because of the complex structure of the metric. For simplicity we consider only the case $E = 0$. As the solution still is too lengthy to be printed here, we will only consider a few results.

For better illustration, we will plot the real part of the quantum fluctuation of the coordinate r on a constant radius as a function of the angle θ (Fig. 22). We see (c.f. Fig. 22) that only at the poles there is a significant distortion of the radius coordinate. In reality (with the correct parameters for the electron for instance) the pole-only effect is even more pronounced. Here we had to resort to less extreme parameters in order to make the spatial distortion visible at all. With increasing distance r this distortion is getting smaller (Fig. 23).

Figure 22 Spatial distortion of the radius coordinate caused by the presence of a charged lepton. A perfect sphere would be obtained in the case of non-distorted flat space-time. In order to make the distortion visible, the following parameters have been applied $r_s = 1, \alpha = 100, r_Q = 20$. The distortion is shown for position $r = r_s$.

Figure 23 Spatial distortion of the radius coordinate caused by the presence of a charged lepton. A perfect sphere would be obtained in the case of non-distorted flat space-time. In order to make the distortion visible, the following parameters have been applied $r_s = 1, \alpha = 100, r_Q = 20$. The distortion is shown for positions $r = r_s$ (yellow), $2r_s, 3r_s, 4r_s$ (red).

Figure 24 Spatial distortion of the radius coordinate caused by the presence of an uncharged lepton. A perfect sphere would be obtained in the case of non-distorted flat space-time. In order to make the distortion visible, the following parameters have been applied $r_s = 1$, $\alpha = 100$, $r_Q = 0$. The distortion is shown for position $r = r_s$.

Interestingly the pole anomaly gets more pronounced in the case of uncharged particles (neutrinos). Figures 24 to 26 give some illustration.

Conclusions to the Quantized Schwarzschild and Kerr Objects

It was shown how the assumption of a substructured space-time automatically leads to a quantum relativity. Starting from an intelligent zero for the line element and a slightly adapted Dirac approach defining the Dirac operator on a metric, we simply have to choose this metric in accordance with the Einstein field equations in order to bring the quantum theory and the general theory of relativity closer together.

With the application of a Schwarzschild metric we obtained an interesting asymmetry between matter and antimatter which might explain the observed dominance of matter over antimatter in our universe.

Figure 25 Spatial distortion of the radius coordinate caused by the presence of an uncharged lepton. A perfect sphere would be obtained in the case of non-distorted flat space-time. In order to make the distortion visible, the following parameters have been applied $r_s = 1$, $\alpha = 100$, $r_Q = 0$. The distortion is shown for positions $r = r_s$ (yellow), $2r_s$, $3r_s$, $4r_s$ (red).

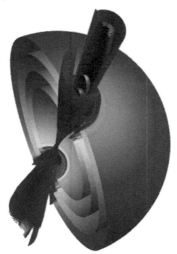

Figure 26 Spatial distortion of the radius coordinate caused by the presence of an uncharged lepton. We have presented again the particle of Fig. 25 with different view angle in order to allow a better observation of the pole region.

Application of a Kerr–Newman–de Sitter metric to the Leptons lead to further solutions for masses, which might have to do with the neutrino masses. Further it was shown that such a non-zero neutrino mass requires a non-zero cosmological constant. The neutrino masses being derived here are extremely small.

Chapter 9

The Photon

The Photon Metric

It is clear from the results of the previous section about the quantized Kerr metric that a potential metric for the photon must have shear components in order to produce wave-like solutions, because we start with the general assumption that the photon "somehow" is a wave.

Therefore, we chose a metric of the form

$$g^{ij} = \begin{pmatrix} g^{00} & 0 & 0 & g^{30} \\ 0 & g^{11} & 0 & 0 \\ 0 & 0 & g^{22} & 0 \\ g^{30} & 0 & 0 & g^{33} \end{pmatrix} \tag{237}$$

where all components are assumed to be constants. It can be shown easily that such a metric always fulfills the Einstein field equations.

The Theory of Everything: Quantum and Relativity is Everywhere – A Fermat Universe
Norbert Schwarzer
Copyright © 2020 Jenny Stanford Publishing Pte. Ltd.
ISBN 978-981-4774-47-5 (Hardcover), 978-1-315-09975-0 (eBook)
www.jennystanford.com

This is leading us to a set of partial differential equations (c.f [9, 10] or previous section) for the component $f = f_{(0)}$ as follows:

$$V_\Omega = \begin{cases} \alpha + \beta + (\gamma + \delta) + a + \overbrace{b + c}^{= 0} + d, \, \alpha + \beta + i\,(\gamma - \delta) + a + b + c + d, \\ \alpha + \beta - i\,(\gamma - \delta) + a + b + c + d, \, \alpha + \beta - (\gamma + \delta) + a + b + c + d, \\ \alpha - \beta + (\gamma + \delta) + a + b + c + d, \, \alpha - \beta + i\,(\gamma - \delta) + a + b + c + d, \\ \alpha - \beta - i\,(\gamma - \delta) + a + b + c + d, \, \alpha - \beta - (\gamma + \delta) + a + b + c + d, \\ \alpha + \beta + (\gamma + \delta) - (a + b + c + d), \, \alpha + \beta + i\,(\gamma - \delta) - (a + b + c + d), \\ \alpha + \beta - i\,(\gamma - \delta) - (a + b + c + d), \, \alpha + \beta - (\gamma + \delta) - (a + b + c + d), \\ \alpha - \beta + (\gamma + \delta) - (a + b + c + d), \, \alpha - \beta + i\,(\gamma - \delta) - (a + b + c + d), \\ \alpha - \beta - i\,(\gamma - \delta) - (a + b + c + d), \, \alpha - \beta - (\gamma + \delta) - (a + b + c + d) \end{cases} = 0.$$

(238)

Here we have set

$$\alpha = i \cdot \sqrt{g^{00}} \cdot \mathbf{f_0}; \; \beta = i \cdot \sqrt{g^{33}} \cdot \mathbf{f_3}; \; \gamma = i \cdot \sqrt{g^{30}} \cdot \mathbf{f_0}; \; \delta = i \cdot \sqrt{g^{30}} \cdot \mathbf{f_3};$$

$$a = g^{00} \cdot \frac{\partial \mathbf{f_0}}{\partial x^0}; \; b = g^{33} \cdot \frac{\partial \mathbf{f_3}}{\partial x^3}; \; c = g^{30} \cdot \frac{\partial \mathbf{f_3}}{\partial x^0}; \; d = g^{30} \cdot \frac{\partial \mathbf{f_0}}{\partial x^3} \qquad (239)$$

and have—for brevity—ignored the f_3-components with respect to the derivatives. A full solution will be given further below. The solutions are easily found for a particle "at rest," because we assume the functions f_i only to be dependent on time and the x_3 respectively the x^3 coordinate. Fortunately, the metric components of our chosen metric do not depend on these coordinates as all components are assumed to be constants. This leaves the equations quite simple and we obtain functions of the form

$$f_{(0)}\left(x^0, x^3\right); \quad \pm I = \{+, -, i, -i\};$$

$$f \to \text{Function}\left[e^{\pm \frac{\left(i\sqrt{g^{00}} \pm I \cdot \sqrt{g^{30}}\right) \cdot t}{g^{00}}} \cdot C \cdot \left[\frac{g^{00} \cdot x^3 - g^{30} \cdot t}{g^{00}}\right]\right], \qquad (240)$$

where C denotes an arbitrary constant, the expression "Function [...]" stands for arbitrary function of the expression in "[...]" and where we find our expected characteristic wave function in t and x^3 of arbitrary form with the argument $\left[\frac{g^{00} \cdot x^3 - g^{30} \cdot t}{g^{00}}\right]$. In addition, we recognize an exponential part similar to the classical Dirac term. Thus, particular solutions could simply be sin- or cos-waves (with arbitrary wavelengths λ):

$$f\left(t, x^3 = z\right) = e^{\pm \frac{\left(i\sqrt{g^{00}} \pm I \cdot \sqrt{g^{30}}\right) \cdot t}{g^{00}}} \cdot C \cdot \begin{cases} \sin\left[\frac{1}{\lambda} \cdot \left(z - \frac{g^{30} \cdot t}{g^{00}}\right)\right] \\ \cos\left[\frac{1}{\lambda} \cdot \left(z - \frac{g^{30} \cdot t}{g^{00}}\right)\right] \end{cases}. \qquad (241)$$

For the photon, we must have of course

$$\frac{g^{30}}{g^{00}} = c, \tag{242}$$

with c denoting the speed of light in vacuum, where it should be pointed out that for the reason of units we should rather see this as const*c. Thereby const shall just be a constant providing the right unit connection, but here we will take it to be just 1. This brings our general solution to

$$f \rightarrow \text{Function} \left[e^{\pm \left(\frac{i}{\sqrt{g^{00}}} \pm \frac{I \cdot c}{\sqrt{g^{30}}} \right) \cdot t} \cdot C \cdot \left[x^3 - c \cdot t \right] \right]. \tag{243}$$

In order to allow only for solutions stable in time, we have to sort out the degrading or infinitely growing solutions $I = \{+, -\}$ and only keep $I = \{+i, -i\}$. This gives us the most general solution for all possible directions of photonic wave propagation in the following form:

$$f \rightarrow \text{Function} \left[e^{\pm \left(\frac{1}{\sqrt{g^{00}}} \pm \frac{c}{\sqrt{g^{j0}}} \right) \cdot i \cdot t} \cdot C \cdot \left[x^j - c \cdot t \right] \right]; \quad j = \{1, 2, 3\}. \tag{244}$$

For completeness, we also give the degrading and singular solutions:

$$f \rightarrow \text{Function} \left[e^{\pm \left(\frac{i}{\sqrt{g^{00}}} \pm \frac{c}{\sqrt{g^{j0}}} \right) \cdot t} \cdot C \cdot \left[x^j - c \cdot t \right] \right]; \quad j = \{1, 2, 3\}. \tag{245}$$

Going back to our example metric (237) we find another set of equations with shear components and the component g^{33} connected with the function f_3. With respect to this function f_3 we obtain the following solution:

$$f_{(3)} \left(x^0, x^3 \right) = F \left(t, x^3 \right); \quad \pm I = \{+, -, i, -i\};$$

$$F \rightarrow \text{Function} \left[e^{\pm \frac{\left(i \sqrt{g^{33}} \pm I \cdot \sqrt{g^{30}} \right) \cdot t}{g^{30}}} \cdot C \cdot \left[x^3 - \frac{g^{33} \cdot t}{g^{30}} \right] \right]. \tag{246}$$

Taking the same arguments as before, this time we have to set

$$\frac{g^{33}}{g^{30}} = c, \tag{247}$$

which would give us the following general solution:

$$F \rightarrow \text{Function} \left[e^{\pm \left(\frac{1}{\sqrt{g^{j0}}} \pm \frac{c}{\sqrt{g^{jj}}} \right) \cdot i \cdot t} \cdot C \cdot \left[x^j - c \cdot t \right] \right]; \quad j = \{1, 2, 3\}. \tag{248}$$

Because they would give wave solutions degenerating (degrade or becoming singular) in time, we have sorted out the following solutions:

$$F \to \text{Function} \left[e^{\left(\pm \frac{1}{\sqrt{g^{j0}}} \pm \frac{c \cdot i}{\sqrt{g^{jj}}} \right) \cdot t} \cdot C \cdot \left[x^j - c \cdot t \right] \right]; \quad j = \{1, 2, 3\}. \quad (249)$$

By the means of virtual parameters as introduced e.g. in (13) or in a more general form in [16], the equations (238) can easily be made eigenvalue equations with respect to certain eigenvalues E. Their solutions would be

$$f \to \text{Function} \left[e^{\left(\pm \frac{i}{\sqrt{g^{00}}} \pm I \cdot \frac{c}{\sqrt{g^{j0}}} \pm e^{\frac{i k \pi}{P}} E_P \right) \cdot t} \cdot C \cdot \left[x^j - c \cdot t \right] \right];$$

$$j = \{1, 2, 3\}; \quad P = \{2, 3, 4, \dots\}; \quad k = \{0, 1, \dots, P - 1\} \quad (250)$$

and

$$F \to \text{Function} \left[e^{\left(\pm \frac{I}{\sqrt{g^{j0}}} \pm \frac{c \cdot i}{\sqrt{g^{jj}}} \pm e^{\frac{i k \pi}{P}} E_P \right) \cdot t} \cdot C \cdot \left[x^j - c \cdot t \right] \right];$$

$$j = \{1, 2, 3\}; \quad P = \{2, 3, 4, \dots\}; \quad k = \{0, 1, \dots, P - 1\} \quad (251)$$

with the corresponding metrics:

$$g^{ij} = \left\{ \begin{pmatrix} g^{00} & 0 & 0 & g^{30} \\ 0 & g^{11} & 0 & 0 \\ 0 & 0 & g^{22} & 0 \\ g^{30} & 0 & 0 & g^{33} \end{pmatrix}, \begin{pmatrix} g^{00} & 0 & g^{20} & 0 \\ 0 & g^{11} & 0 & 0 \\ g^{20} & 0 & g^{22} & 0 \\ 0 & 0 & 0 & g^{33} \end{pmatrix}, \begin{pmatrix} g^{00} & g^{10} & 0 & 0 \\ g^{10} & g^{11} & 0 & 0 \\ 0 & 0 & g^{22} & 0 \\ 0 & 0 & 0 & g^{33} \end{pmatrix} \right\}.$$

$$(252)$$

Alternatively, we could apply the relations (242) and (247) in order to substitute one of the metric components, giving us

$$f \to \text{Function} \left[e^{\left(\frac{\sqrt{c}}{\sqrt{g^{j0}}} (\pm i \pm I \sqrt{c}) \pm e^{\frac{i k \pi}{P}} E_P \right) \cdot t} \cdot C \cdot \left[x^j - c \cdot t \right] \right]; \quad c = \frac{g^{j0}}{g^{00}}$$

$$j = \{1, 2, 3\}; \quad P = \{2, 3, 4, \dots\}; \quad k = \{0, 1, \dots, P - 1\} \quad (253)$$

and

$$F \to \text{Function} \left[e^{\left(\pm \frac{I}{\sqrt{g^{j0}}} \pm \frac{\sqrt{c} \cdot i}{\sqrt{g^{j0}}} \pm e^{\frac{i k \pi}{P}} E_P \right) \cdot t} \cdot C \cdot \left[x^j - c \cdot t \right] \right]; \quad c = \frac{g^{jj}}{g^{j0}}.$$

$$j = \{1, 2, 3\}; \quad P = \{2, 3, 4, \dots\}; \quad k = \{0, 1, \dots, P - 1\} \quad (254)$$

Further generalization with respect to arbitrary directions of wave propagation and introduction of the wave-vector $\hat{\mathbf{k}}$ results in

$$f \rightarrow \text{Function}\left[e^{\left(\frac{\sqrt{c}}{\sqrt{g^{j0}}}(\pm i \pm I\sqrt{c}) \pm e^{\frac{i k_{\pi}}{P}} E_P \right)\cdot t} \cdot \vec{\mathbf{C}}_f \cdot \left[\hat{\mathbf{k}}_f \cdot \mathbf{x} - c \cdot t \right] \right]; \quad c = \frac{g^{j0}}{g^{00}}$$

$$j = \{1, 2, 3\}; \quad P = \{2, 3, 4, \ldots\}; \quad k = \{0, 1, \ldots, P-1\} \tag{255}$$

and

$$F \rightarrow \text{Function}\left[e^{\left(\pm \frac{I}{\sqrt{g^{j0}}} \pm \frac{\sqrt{c}\cdot i}{\sqrt{g^{j0}}} \pm e^{\frac{i k_{\pi}}{P}} E_P \right)\cdot t} \cdot \vec{\mathbf{C}}_F \cdot \left[\hat{\mathbf{k}}_F \cdot \mathbf{x} - c \cdot t \right] \right]; \quad c = \frac{g^{jj}}{g^{j0}}$$

$$j = \{1, 2, 3\}; \quad P = \{2, 3, 4, \ldots\}; \quad k = \{0, 1, \ldots, P-1\}. \tag{256}$$

Superposition of the wave-solutions above and various choices for the "Function [. . .]" option provide a great variety of metric waves.

Connections with Maxwell

For further consideration, we need the following derivatives of our general solutions:

$$\left\{ \vec{\nabla}\times, \vec{\nabla}, \partial_t \right\} f \rightarrow e^{\chi \cdot t} \cdot \left\{ \hat{\mathbf{k}}_f \times \vec{\mathbf{C}}_f \cdot A'_f, \; \hat{\mathbf{k}}_f \cdot \vec{\mathbf{C}}_f \cdot A'_f, \; \chi \cdot \vec{\mathbf{C}}_f \cdot A_f - c \cdot \vec{\mathbf{C}}_f \cdot A'_f \right\}$$

$$A'_f = \frac{\partial \text{Function}\left[\hat{\mathbf{k}}_f \cdot \mathbf{x} - c \cdot t \right]}{\partial \left(\hat{\mathbf{k}}_f \cdot \mathbf{x} - c \cdot t \right)}; \quad A_f = \text{Function}\left[\hat{\mathbf{k}}_f \cdot \mathbf{x} - c \cdot t \right];$$

$$\chi = \left(\frac{\sqrt{c}}{\sqrt{g^{j0}}} (\pm i \pm I\sqrt{c}) \pm e^{\frac{i k_{\pi}}{P}} E_P \right) \tag{257}$$

and

$$\left\{ \vec{\nabla}\times, \vec{\nabla}, \partial_t \right\} F \rightarrow e^{\kappa \cdot t} \cdot \left\{ \hat{k}_F \times \vec{\mathbf{C}}_F \cdot A'_F, \; \hat{k}_F \cdot \vec{\mathbf{C}}_F \cdot A'_F, \; \kappa \cdot \vec{\mathbf{C}}_F \cdot A_F - c \cdot \vec{\mathbf{C}}_F \cdot A'_F \right\}$$

$$A'_F = \frac{\partial \text{Function}\left[\hat{k}_F \cdot \mathbf{x} - c \cdot t \right]}{\partial \left(\hat{k}_F \cdot \mathbf{x} - c \cdot t \right)}; \quad A_F = \text{Function}\left[\hat{k}_F \cdot \mathbf{x} - c \cdot t \right];$$

$$\kappa = \left(\pm \frac{I}{\sqrt{g^{j0}}} \pm \frac{\sqrt{c}\cdot i}{\sqrt{g^{j0}}} \pm e^{\frac{i k_{\pi}}{P}} E_P \right). \tag{258}$$

Now we demand the divergence to be zero (no sources) for both types of solutions, which gives us

$$\hat{\mathbf{k}}_F \cdot \vec{\mathbf{C}}_F = 0 = \hat{\mathbf{k}}_f \cdot \vec{\mathbf{C}}_f, \tag{259}$$

which means that the fields building the waves are orthogonal to the direction of propagation. This, of course, is motivated by the first and second Maxwell equation and nurtured by the suspicion that the two solutions f and F have to do with the electric and the magnetic field.

The second condition we might want to set would be motivated by the third and the fourth of the Maxwell equations. It combines the rotation and the time derivative of the two waves and gives

$$\vec{\nabla} \times f = -\partial_t F \rightarrow e^{\chi \cdot t} \cdot \hat{k}_f \times \vec{C}_f \cdot A'_f = -e^{\kappa \cdot t} \cdot \left(\kappa \cdot \vec{C}_F \cdot A_F - c \cdot \vec{C}_F \cdot A'_F \right)$$

$$\vec{\nabla} \times F = \mu_0 \vec{j} + \frac{\partial_t f}{c^2} \rightarrow e^{\kappa \cdot t} \cdot \hat{k}_F \times \vec{C}_F \cdot A'_F$$

$$= \mu_0 \vec{j} + \frac{e^{\chi \cdot t}}{c^2} \cdot \left(\chi \cdot \vec{C}_f \cdot A_f - c \cdot \vec{C}_f \cdot A'_f \right). \tag{260}$$

Here $\vec{j}$ and μ_0 denote the displacement current density and the permeability, respectively. We will not need the fourth condition here, but want to discuss it later on. We find that the conditions above require the exponents in the e-function to be zero, which requires different eigenvalues E_{Pf} and E_{PF} for the two waves:

$$\chi = \left(\frac{\sqrt{c}}{\sqrt{g^{j0}}} \left(\pm i \pm I \sqrt{c} \right) \pm e^{\frac{i k_\pi}{P}} E_{Pf} \right) = 0$$

$$= \left(\pm \frac{I}{\sqrt{g^{j0}}} \pm \frac{\sqrt{c} \cdot i}{\sqrt{g^{j0}}} \pm e^{\frac{i k_\pi}{P}} E_{PF} \right) \doteq \kappa. \tag{261}$$

This simplifies equation (260) to

$$\vec{\nabla} \times f = -\partial_t F \rightarrow \hat{k}_f \times \vec{C}_f \cdot A'_f = c \cdot \vec{C}_F \cdot A'_F$$

$$\vec{\nabla} \times F = \mu_0 \vec{j} + \frac{\partial_t f}{c^2} \rightarrow \hat{k}_F \times \vec{C}_F \cdot A'_F = \mu_0 \vec{j} - \frac{\vec{C}_f \cdot A'_f}{c}. \tag{262}$$

The first equation requires the wave vectors to be equal:

$$\hat{k}_F = \hat{k}_f \tag{263}$$

and fixes the $\vec{C}_F, \vec{C}_f$ such that the F-wave is standing orthogonal on f, because we have

$$\vec{\nabla} \times f = -\partial_t F \rightarrow \hat{k}_f \times \vec{C}_f = c \cdot \vec{C}_F. \tag{264}$$

Now we easily recognize the following identities for the electric $\vec{E}$ and the magnetic field $\vec{B}$:

$$\vec{E} = \vec{C}_f; \quad \vec{B} = \vec{C}_F. \tag{265}$$

The Other Way to Fulfill the Maxwell Equations with Plane Waves

An apparently simpler way to fulfill the Maxwell equations with our solutions at hand would be the following:

We allow the wave function part "Function [. . .]" to be complex, thus, forming a two component vector. This could simply be achieved by considering the vectors $\vec{\mathbf{C}}_F$, $\vec{\mathbf{C}}_f$ to be complex. Equation (259) now needs to be fulfilled for the real Re [. . .] and the imaginary Im [. . .] parts of the C vectors separately, leaving us with

$$\hat{\mathbf{k}}_F \cdot \text{Re}\left[\vec{\mathbf{C}}_F\right] = \hat{\mathbf{k}}_F \cdot \text{Im}\left[\vec{\mathbf{C}}_F\right] = 0 = \hat{\mathbf{k}}_f \cdot \text{Im}\left[\vec{\mathbf{C}}_f\right] = \hat{\mathbf{k}}_f \cdot \text{Re}\left[\vec{\mathbf{C}}_f\right]. \quad (266)$$

Then we demand the first condition in (262) as follows for f and F:

$$\vec{\nabla} \times \text{Re}\,[f] = -\partial_t \text{Im}\,[f]$$
$$\rightarrow \hat{k}_f \times \text{Re}\left[e^{\chi \cdot t} \cdot \vec{\mathbf{C}}_f\right] \cdot A'_f = -\text{Im}\left[e^{\chi \cdot t} \cdot \left(\chi \cdot \vec{\mathbf{C}}_f \cdot A_f - c \cdot \vec{\mathbf{C}}_f \cdot A'_f\right)\right]$$
$$\vec{\nabla} \times \text{Re}\,[F] = -\partial_t \text{Im}\,[F]$$
$$\rightarrow \hat{k}_F \times \text{Re}\left[e^{\kappa \cdot t} \cdot \vec{\mathbf{C}}_F\right] \cdot A'_F = -\text{Im}\left[e^{\kappa \cdot t} \cdot \left(\kappa \cdot \vec{\mathbf{C}}_F \cdot A_F \quad c \cdot \vec{\mathbf{C}}_F \cdot A'_F\right)\right],$$
$$(267)$$

which again demands the exponents χ, κ to be zero and results in

$$\begin{aligned} \hat{\mathbf{k}}_f \times \text{Re}\left[\vec{\mathbf{C}}_f\right] = c \cdot \text{Im}\left[\vec{\mathbf{C}}_f\right] \\ \hat{\mathbf{k}}_F \times \text{Re}\left[\vec{\mathbf{C}}_F\right] = c \cdot \text{Im}\left[\vec{\mathbf{C}}_F\right] \end{aligned} \quad (268)$$

Now we can have independent $\hat{\mathbf{k}}_F$, $\hat{\mathbf{k}}_f$ and therefore obtain two solutions compatible with the Maxwell conditions. The relationship with the electromagnetic field is given as

$$\vec{\mathbf{E}}_1 = \text{Re}\left[\vec{\mathbf{C}}_f\right] \text{ and } \vec{\mathbf{E}}_2 = \text{Re}\left[\vec{\mathbf{C}}_F\right]; \ \vec{\mathbf{B}}_1 = \text{Im}\left[\vec{\mathbf{C}}_f\right] \text{ and } \vec{\mathbf{B}}_2 = \text{Im}\left[\vec{\mathbf{C}}_F\right]. \quad (269)$$

Interpreting, as Dirac has done, the solutions as particles, one immediately can see that the condition $\chi = \kappa = 0$ demands the particles to be massless, because comparison with the Dirac solutions for particles at rest (c.f. [5]) shows that the exponent in the exponential function contains the mass M of the particle. For convenience, we here give the Dirac solution as limit of the quantized Schwarzschild solution $f_{(0)}$ (Schwarzschild particle at rest, see

[17]) for great distances r from the particle's center position:

$$f_{(0)}(t, r) = \{C_{01} \cdot g_1(t), C_{02} \cdot g_1(t), C_{03} \cdot g_2(t), C_{04} \cdot g_2(t)\}$$

$$\lim_{r \to \infty} g_1 = \lim_{r \to \infty} e^{-\frac{i c^2 \left(e^{\frac{i \cdot j \cdot \pi}{P}} E_P + \mu \sqrt{\frac{r}{c^2(r-r_s)}}\right) \cdot (r-r_s) \cdot t}{r}}$$

$$= e^{-i \cdot c^2 t \left(e^{\frac{i \cdot j \cdot \pi}{P}} E_P + \frac{\mu}{c}\right)} \quad \underrightarrow{E_P = 0, \ \text{Dirac}} \quad e^{-i \cdot c \cdot t \cdot \mu}$$

$$\lim_{r \to \infty} g_2 = \lim_{r \to \infty} e^{\frac{i c^2 \left(e^{\frac{i \cdot j \cdot \pi}{P}} E_P + \mu \sqrt{\frac{r}{c^2(r-r_s)}}\right) \cdot (r-r_s) \cdot t}{r}}$$

$$= e^{i \cdot c^2 t \left(e^{\frac{i \cdot j \cdot \pi}{P}} E_P + \frac{\mu}{c}\right)} \quad \underrightarrow{E_P = 0, \ \text{Dirac}} \quad e^{i \cdot c \cdot t \cdot \mu}, \tag{270}$$

with the parameter μ we have the mass M in connection with the speed of light and the Planck constant $\hbar$:

$$\mu = \frac{c \cdot M}{\hbar}. \tag{271}$$

Thus, as said before, our wave solutions given above in (255) and (256), in combination with the Maxwell boundary conditions, lead to massless particles. We might as well interpret them as photons.

The other possible solutions are not massless and thus, are probably connected with the weak interaction.

Illustrations

In order to illustrate and better understand the nature of the solutions found, we need to apply (52) and to evaluate the base vectors of the photonic metric. We pick (237) as an example and find the base vector solutions by solving the following system of equations:

$$\mathbf{e}_0 = \begin{pmatrix} E_0 \\ 0 \\ 0 \\ E_{30} \end{pmatrix}; \quad \mathbf{e}_1 = \begin{pmatrix} 0 \\ E_1 \\ 0 \\ 0 \end{pmatrix}; \quad \mathbf{e}_1 = \begin{pmatrix} 0 \\ 0 \\ E_2 \\ 0 \end{pmatrix}; \quad \mathbf{e}_3 = \begin{pmatrix} E_{30} \\ 0 \\ 0 \\ E_3 \end{pmatrix}$$

$$\mathbf{e}_0 \cdot \mathbf{e}_0 = \gamma_{00}; \quad \mathbf{e}_1 \cdot \mathbf{e}_1 = \gamma_{11}; \quad \mathbf{e}_2 \cdot \mathbf{e}_2 = \gamma_{22}; \quad \mathbf{e}_3 \cdot \mathbf{e}_3 = \gamma_{33}; \quad \mathbf{e}_0 \cdot \mathbf{e}_3 = \gamma_{30} \tag{272}$$

and similarly for the contravariant basis. The solutions are simple but lengthy. Thus, we will not give them here.

We find that according to equation (52) and caused by the shear elements in the quantum part this automatically connects time and space coordinates as follows:

$$\vec{t} = \mathbf{x_0} = \begin{pmatrix} G_0 \\ 0 \\ 0 \\ 0 \end{pmatrix} + \begin{pmatrix} E_0 \\ 0 \\ 0 \\ E_{30} \end{pmatrix} f_0\,(t,z); \quad \vec{x} = \mathbf{x_1} = \begin{pmatrix} 0 \\ G_1 \\ 0 \\ 0 \end{pmatrix} + \begin{pmatrix} 0 \\ E_1 \\ 0 \\ 0 \end{pmatrix} f_1\,(x);$$

$$\vec{y} = \mathbf{x_2} = \begin{pmatrix} 0 \\ 0 \\ G_2 \\ 0 \end{pmatrix} + \begin{pmatrix} 0 \\ 0 \\ E_2 \\ 0 \end{pmatrix} f_2\,(y); \quad \vec{z} = \mathbf{x_3} = \begin{pmatrix} 0 \\ 0 \\ 0 \\ G_3 \end{pmatrix} + \begin{pmatrix} E_{30} \\ 0 \\ 0 \\ E_3 \end{pmatrix} f_3\,(t,z).$$

$$(273)$$

While in the case of plane waves, for x and y only the usual quantum jitter with f_1 and f_2 (c.f. [9], second example) is obtained, we have a space-time connection for the t and z components. For better illustration, we have separated the classical metric base vectors denoted with G_i from the quantum mechanical ones denoted with E_i. We see that in the presence of a particle with metric shear components and photonic properties the quantum jitter does not leave the coordinates independent anymore, but distorts them such, that the two coordinates being sheared against each other are connected. This connection propagates like a wave with the speed of light c.

For optimum presentation, we will also consider here wave packages of Gaussian shape instead of infinite plane waves. Such packages are easily obtained by superposing plane waves with amplitudes of the form

$$\vec{\mathbf{C}}_f = \vec{\mathbf{C}}_{f0} \cdot C\,(k); \quad \vec{\mathbf{C}}_F = \vec{\mathbf{C}}_{F0} \cdot C\,(k); \quad C\,(k) = e^{-\frac{(k-k_0)^2 a^2}{2}}, \qquad (274)$$

with a determining the width of the wave package.

Figure 27 gives an illustrative example for such a "Gaussian photon" traveling in z-direction. It is—unrealistically—assumed that the particle only exists in t and the direction of propagation, which is to say it is two-dimensional. We will later see how to make this more realistic.

Spatial Extension of the Solution and the Localized Photon

It is clear from the Maxwell condition (259) that for a real photon also the other spatial axes x and y must be influenced respectively distorted.

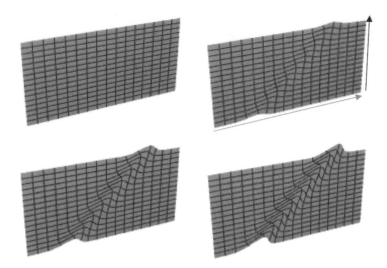

Figure 27 Illustration of the time-space disturbance caused by a passing photon. While the upper left figure shows no photon, the other three pictures give a demonstration of various "Gaussian photons" (wave packages of Gaussian shape; see text) with increasing strength. The blue arrow gives the z axis, the black one stands for time.

Thus, any photon traveling in direction of z would not only show distortion along the z axis and with respect to time, but also in x and y. Not unlike an elastic body with non-zero Poisson's ratio we should expect secondary distortions along the axes perpendicular to the axis of main distortion. On the other hand, such a distortion should decrease with the distance of the localized source. Thus, for a truly particle-like photon, one would not only expect all spatial axes to be somehow influenced, but should also observe convergence with respect to all coordinates. Thus, at first we perform the following superposition of our plane wave solutions:

$$f = e^{\left(\frac{\sqrt{c}}{\sqrt{g^{j0}}}\left(\pm i \pm I \sqrt{c}\right) \pm e^{\frac{i \mathbf{k}_\pi}{P}} E_P\right) \cdot t} \cdot \vec{\mathbf{C}}_f \cdot \frac{\sigma}{\sqrt{2\pi}} \int\limits_{-\infty}^{\infty} e^{-\frac{\sigma^2\left(\alpha - \alpha_{f0}\right)^2}{2}} \cdot e^{i \cdot \alpha \cdot \left[\hat{\mathbf{k}}_f \cdot \mathbf{x} - c \cdot t\right]} d\alpha;$$

$$c = \frac{g^{j0}}{g^{00}} \qquad j = \{1, 2, 3\}; \qquad P = \{2, 3, 4, \ldots\}; \qquad k = \{0, 1, \ldots, P-1\}$$

$$(275)$$

and

$$F = e^{\left(\pm\frac{I}{\sqrt{g^{j0}}}\pm\frac{\sqrt{c}\cdot i}{\sqrt{g^{j0}}}\pm e^{\frac{i\cdot k\cdot\pi}{P}}E_P\right)\cdot t} \cdot \vec{C}_F \cdot \frac{\sigma}{\sqrt{2\pi}} \int\limits_{-\infty}^{\infty} e^{-\frac{\sigma^2(\alpha-\alpha_{F0})^2}{2}} \cdot e^{i\cdot\alpha\cdot\left[\hat{k}_F\cdot x - c\cdot t\right]} d\alpha;$$

$$c = \frac{g^{jj}}{g^{j0}} \quad j = \{1, 2, 3\}; \quad P = \{2, 3, 4, \ldots\}; \quad k = \{0, 1, \ldots, P-1\}.$$

(276)

With this we obtain localized solutions with respect to the z coordinate

$$f = e^{\left(\frac{\sqrt{c}}{\sqrt{g^{j0}}}(\pm i \pm I\sqrt{c})\pm e^{\frac{i\cdot k\cdot\pi}{P}}E_P\right)\cdot t} \cdot \vec{C}_f \cdot e^{i\cdot\alpha_{f0}\cdot\left[\hat{k}_f\cdot x - c\cdot t\right] - \frac{1}{2\sigma^2}\left[\hat{k}_f\cdot x - c\cdot t\right]^2}$$

(277)

and

$$F = e^{\left(\pm\frac{I}{\sqrt{g^{j0}}}\pm\frac{\sqrt{c}\cdot i}{\sqrt{g^{j0}}}\pm e^{\frac{i\cdot k\cdot\pi}{P}}E_P\right)\cdot t} \cdot \vec{C}_F \cdot e^{i\cdot\alpha_{F0}\cdot\left[\hat{k}_F\cdot x - c\cdot t\right] - \frac{1}{2\sigma^2}\left[\hat{k}_F\cdot x - c\cdot t\right]^2}.$$

(278)

In the same way we superpose the quantum jitter solutions f_1 and f_2 for x and y:

$$f_1 = \vec{C}_x \cdot \frac{\sigma\sqrt{\sqrt{g^{11}}}}{\sqrt{2\pi}} \int\limits_{-\infty}^{\infty} e^{-\frac{\sigma^2(\alpha-\alpha_{x0})^2}{2\sqrt{g^{11}}}} \cdot e^{\frac{i\cdot\alpha\cdot x\pm e^{\frac{i\cdot k\cdot\pi}{P}}E_{Px}\cdot x}{\sqrt{g^{11}}}} d\alpha$$

$$= \vec{C}_x \cdot e^{\frac{\pm i\cdot\alpha_{x0}x\pm e^{\frac{i\cdot k\cdot\pi}{P}}E_{Px}\cdot x - \frac{x^2}{2\cdot\sigma^2}}{\sqrt{g^{11}}}}$$

$$f_2 = \vec{C}_y \cdot \frac{\sigma\sqrt{\sqrt{g^{22}}}}{\sqrt{2\pi}} \int\limits_{-\infty}^{\infty} e^{-\frac{\sigma^2(\alpha-\alpha_{y0})^2}{2}} \cdot e^{\frac{i\cdot\alpha\cdot y\pm e^{\frac{i\cdot k\cdot\pi}{P}}E_{Py}\cdot y}{\sqrt{g^{22}}}} d\alpha$$

$$= \vec{C}_y \cdot e^{\frac{\pm i\cdot\alpha_{y0}y\pm e^{\frac{i\cdot k\cdot\pi}{P}}E_{Py}\cdot y - \frac{y^2}{2\cdot\sigma^2}}{\sqrt{g^{22}}}}$$

$$j = \{1, 2, 3\}; \quad P = \{2, 3, 4, \ldots\}; \quad k = \{0, 1, \ldots, P-1\}. \quad (279)$$

We should point out here that for generality different $\sigma \to \sigma_x, \sigma_y, \sigma_z$ could be demanded for the various axes x, y, and z. For simplicity and brevity, we will not incorporate this additional degree of freedom here. For simplicity, we set $\alpha_{x0} = \alpha_{y0} = E_{Px} = E_{Py} = 0$ with respect to our further considerations, which is just sorting out the usual plane wave quantum jitter and further options for superposition. Later it could be added in again if of need. Keeping the condition $\chi = \kappa = 0$ and fixing the propagation vector in the z direction, we seek the total spatial solution for the localized photon in the form

$$F_{xyz} = f_1 \cdot f_2 \cdot F = c_x \cdot c_y \cdot \vec{C}_F \cdot e^{-\frac{x^2}{2 \cdot \sigma^2 \sqrt{g^{11}}} - \frac{y^2}{2 \cdot \sigma^2 \sqrt{g^{22}}} + i \cdot \alpha_{F0} \cdot [k_z \cdot z - c \cdot t] - \frac{1}{2\sigma^2} \cdot [k_z \cdot z - c \cdot t]^2}$$

$$= \vec{C}_{xyz} \cdot e^{-\frac{x^2}{2 \cdot \sigma^2 \sqrt{g^{11}}} - \frac{y^2}{2 \cdot \sigma^2 \sqrt{g^{22}}} + i \cdot \alpha_{F0} \cdot [k_z \cdot z - c \cdot t] - \frac{1}{2\sigma^2} \cdot [k_z \cdot z - c \cdot t]^2}. \tag{280}$$

This solution still fulfills all differential equations resulting from the metric (237), but we are getting some problems with the Maxwell conditions. Again, we evaluate the following derivatives:

$$\left\{ \vec{\nabla} \times, \vec{\nabla} \cdot, \partial_t \right\} F_{xyz} \rightarrow e^{\beta} \cdot \left\{ \vec{\nabla} \times \vec{C}_{xyz} \beta, \vec{\nabla} \beta \cdot \vec{C}_{xyz}, \partial_t \beta \cdot \vec{C}_{xyz} \right\}$$

$$\beta = -\frac{x^2}{2 \cdot \sigma^2 \sqrt{g^{11}}} - \frac{y^2}{2 \cdot \sigma^2 \sqrt{g^{22}}} + i \cdot \alpha_{F0} \cdot [k_z \cdot z - c \cdot t] - \frac{1}{2\sigma^2} \cdot [k_z \cdot z - c \cdot t]^2$$

$$\vec{\nabla} \times \vec{C}_{xyz} \beta = \beta \cdot \vec{\nabla} \times \vec{C}_{xyz} - \frac{x}{\sigma^2 \sqrt{g^{11}}} \begin{pmatrix} 0 \\ -\tilde{C}_z \\ \tilde{C}_y \end{pmatrix} - \frac{y}{\sigma^2 \sqrt{g^{22}}} \begin{pmatrix} \tilde{C}_z \\ 0 \\ -\tilde{C}_x \end{pmatrix}$$

$$+ \left(i \cdot \alpha_{F0} \cdot k_z - \frac{k_z}{\sigma^2} \cdot [k_z \cdot z - c \cdot t] \right) \begin{pmatrix} -\tilde{C}_y \\ \tilde{C}_x \\ 0 \end{pmatrix}$$

$$\vec{\nabla} \beta = -\frac{x}{\sigma^2 \sqrt{g^{11}}} \begin{pmatrix} 1 \\ 0 \\ 0 \end{pmatrix} - \frac{y}{\sigma^2 \sqrt{g^{22}}} \begin{pmatrix} 0 \\ 1 \\ 0 \end{pmatrix}$$

$$+ \left(i \cdot \alpha_{F0} \cdot k_z - \frac{k_z}{\sigma^2} \cdot [k_z \cdot z - c \cdot t] \right) \begin{pmatrix} 0 \\ 0 \\ 1 \end{pmatrix}$$

$$\partial_t \beta = -i \cdot c \cdot \alpha_{F0} + \frac{c}{\sigma^2} \cdot [k_z \cdot z - c \cdot t]; \quad \vec{C}_{xyz} = \begin{pmatrix} \tilde{C}_x \\ \tilde{C}_y \\ \tilde{C}_z \end{pmatrix}. \tag{281}$$

One easily sees that all Maxwell conditions can be fulfilled as before in the case of the imaginary part of the solution. Things are a bit different for the real part, however. Concentrating first on the divergence $\vec{\nabla} \beta \cdot \vec{C}_{xyz}$, we find that any vector of the form

$$\text{Re} \left[\vec{C}_F \right] = \begin{pmatrix} \tilde{C}_x = C_x (x, y) \\ -\tilde{C}_y = -C_y (x, y) \\ 0 \end{pmatrix} = \begin{pmatrix} C_x \cdot y \\ -C_x \cdot x \\ 0 \end{pmatrix}$$

$$\Rightarrow \vec{C}_{xyz} = c_x \cdot c_y \cdot \vec{C}_F = c_x \cdot c_y \cdot C_x \begin{pmatrix} y + i \cdot f_x (y) \\ -x + i \cdot f_y (x) \\ 0 \end{pmatrix} \tag{282}$$

with $C_{x,y}(*)$ standing for functions and $C_{x,y}$ denoting constants, would give us the desired result.

Now we perform the same evaluation for the function f:

$$f_{xyz} = f_1 \cdot f_2 \cdot f = c_x \cdot c_y \cdot \vec{\mathbf{C}}_f \cdot e^{-\frac{x^2}{2\cdot\sigma^2\sqrt{g^{11}}} - \frac{y^2}{2\cdot\sigma^2\sqrt{g^{22}}} + i\cdot\alpha_{f0}\cdot[k_z\cdot z - c\cdot t] - \frac{1}{2\sigma^2}\cdot[k_z\cdot z - c\cdot t]^2}$$

$$= \vec{\mathbf{C}}_{fxyz} \cdot e^{-\frac{x^2}{2\cdot\sigma^2\sqrt{g^{11}}} - \frac{y^2}{2\cdot\sigma^2\sqrt{g^{22}}} + i\cdot\alpha_{f0}\cdot[k_z\cdot z - c\cdot t] - \frac{1}{2\sigma^2}\cdot[k_z\cdot z - c\cdot t]^2} \tag{283}$$

which gives us

$$\{\vec{\nabla}\times, \vec{\nabla}\cdot, \partial_t\} f_{xyz} \rightarrow e^{\beta} \cdot \{\vec{\nabla}\times\vec{\mathbf{C}}_{fxyz}\beta, \vec{\nabla}\beta\cdot\vec{\mathbf{C}}_{fxyz}, \partial_t\beta\cdot\vec{\mathbf{C}}_{fxyz}\}$$

$$\beta = -\frac{x^2}{2\cdot\sigma^2\sqrt{g^{11}}} - \frac{y^2}{2\cdot\sigma^2\sqrt{g^{22}}} + i\cdot\alpha_{f0}\cdot[k_z\cdot z - c\cdot t] - \frac{1}{2\sigma^2}\cdot[k_z\cdot z - c\cdot t]^2$$

$$\vec{\nabla}\times\vec{\mathbf{C}}_{fxyz}\beta = \beta\cdot\vec{\nabla}\times\vec{\mathbf{C}}_{fxyz} - \frac{x}{\sigma^2\sqrt{g^{11}}}\begin{pmatrix}0\\-\tilde{C}_z\\\tilde{C}_y\end{pmatrix} - \frac{y}{\sigma^2\sqrt{g^{22}}}\begin{pmatrix}\tilde{C}_z\\0\\-\tilde{C}_x\end{pmatrix}$$

$$+\left(i\cdot\alpha_{f0}\cdot k_z - \frac{k_z}{\sigma^2}\cdot[k_z\cdot z - c\cdot t]\right)\begin{pmatrix}-\tilde{C}_y\\\tilde{C}_x\\0\end{pmatrix}$$

$$\vec{\nabla}\beta = -\frac{x}{\sigma^2\sqrt{g^{11}}}\begin{pmatrix}1\\0\\0\end{pmatrix} - \frac{y}{\sigma^2\sqrt{g^{22}}}\begin{pmatrix}0\\1\\0\end{pmatrix}$$

$$+\left(i\cdot\alpha_{f0}\cdot k_z - \frac{k_z}{\sigma^2}\cdot[k_z\cdot z - c\cdot t]\right)\begin{pmatrix}0\\0\\1\end{pmatrix}$$

$$\partial_t\beta = -i\cdot c\cdot\alpha_{f0} + \frac{c}{\sigma^2}\cdot[k_z\cdot z - c\cdot t]; \quad \vec{\mathbf{C}}_{fxyz} = \begin{pmatrix}\tilde{C}_x\\\tilde{C}_y\\\tilde{C}_z\end{pmatrix}. \tag{284}$$

Now the divergence $\vec{\nabla}\beta\cdot\vec{\mathbf{C}}_{fxyz}$ gives us the vector $\vec{\mathbf{C}}_{fxyz}$

$$\mathrm{Re}\left[\vec{\mathbf{C}}_f\right] = \begin{pmatrix}\tilde{C}_x = C_x\,(x,y)\\-\tilde{C}_y = -C_y\,(x,y)\\0\end{pmatrix} = \begin{pmatrix}C_x\cdot y\\-C_x\cdot x\\0\end{pmatrix}$$

$$\Rightarrow \vec{\mathbf{C}}_{fxyz} = c_x\cdot c_y\cdot\vec{\mathbf{C}}_f = c_x\cdot c_y\cdot C_x\begin{pmatrix}y + i\cdot f_x\,(y)\\-x + i\cdot f_y\,(x)\\0\end{pmatrix} \tag{285}$$

as a possible solution for a divergence-free (solenoidal) localized photon. We would still have to consider the third Maxwell law for both f and F, but at first we want to discuss briefly the funny requirement for the vectors $\vec{\mathbf{C}}_{fxyz}$ and $\vec{\mathbf{C}}_{xyz}$ arising from the divergence $= 0$ condition.

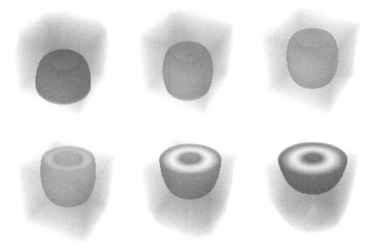

Figure 28 Localized (Gaussian), solenoidal photon passing along the z axis. Absolute field values are shown. One easily recognizes the spin due to the hollow shape of the structure.

Localizing the Photon Forces It to Evolve Spin

We see that the localization of the photon, in connection with the first and second Maxwell law (divergence $= 0$) requires structures of the vector fields such that they end of with spin. Figure 28 shows this as the absolute value field for a passing photon.

It was derived that without mass, which in our case is achieved by the condition $\chi = \kappa = 0$, solenoidality can only be achieved with spin.

For our further considerations, we will also fade out the plane wave in the z direction, which gives us

$$F_{xyz} = c_x \cdot c_y \cdot \vec{\mathbf{C}}_F \cdot e^{-\frac{x^2}{2 \cdot \sigma^2 \sqrt{g^{11}}} - \frac{y^2}{2 \cdot \sigma^2 \sqrt{g^{22}}} - \frac{1}{2\sigma^2} \cdot [k_z \cdot z - c \cdot t]^2}$$

$$= C \cdot \begin{pmatrix} y \\ -x \\ 0 \end{pmatrix} \cdot e^{-\frac{x^2}{2 \cdot \sigma^2 \sqrt{g^{11}}} - \frac{y^2}{2 \cdot \sigma^2 \sqrt{g^{22}}} - \frac{1}{2\sigma^2} \cdot [k_z \cdot z - c \cdot t]^2}, \tag{286}$$

$$f_{xyz} = f_1 \cdot f_2 \cdot f = c_x \cdot c_y \cdot \vec{\mathbf{C}}_f \cdot e^{-\frac{x^2}{2 \cdot \sigma^2 \sqrt{g^{11}}} - \frac{y^2}{2 \cdot \sigma^2 \sqrt{g^{22}}} - \frac{1}{2\sigma^2} \cdot [k_z \cdot z - c \cdot t]^2}$$

$$= \vec{\mathbf{C}}_{fxyz} \cdot e^{-\frac{x^2}{2 \cdot \sigma^2 \sqrt{g^{11}}} - \frac{y^2}{2 \cdot \sigma^2 \sqrt{g^{22}}} - \frac{1}{2\sigma^2} \cdot [k_z \cdot z - c \cdot t]^2} \tag{287}$$

simplifying our derivatives as follows:

$$\left\{ \vec{\nabla} \times, \vec{\nabla} \cdot, \partial_t \right\} F_{xyz} \rightarrow e^{\beta} \cdot \left\{ \vec{\nabla} \times \vec{C}_{xyz} + \vec{\nabla}\beta \times \vec{C}_{xyz}, \vec{\nabla}\beta \cdot \vec{C}_{xyz}, \partial_t\beta \cdot \vec{C}_{xyz} \right\}$$

$$\beta = -\frac{x^2}{2 \cdot \sigma^2 \sqrt{g^{11}}} - \frac{y^2}{2 \cdot \sigma^2 \sqrt{g^{22}}} - \frac{1}{2\sigma^2} \cdot [k_z \cdot z - c \cdot \mathbf{t}]^2$$

$$\vec{\nabla} \times \vec{C}_{xyz} + \vec{\nabla}\beta \times \vec{C}_{xyz} = C \cdot \left(\frac{x}{\sigma^2\sqrt{g^{11}}} \begin{pmatrix} 0 \\ 0 \\ x \end{pmatrix} + \frac{y}{\sigma^2\sqrt{g^{22}}} \begin{pmatrix} 0 \\ 0 \\ y \end{pmatrix} \right.$$

$$\left. - \frac{k_z}{\sigma^2} \cdot [k_z \cdot z - c \cdot \mathbf{t}] \left(\begin{pmatrix} x \\ y \\ 0 \end{pmatrix} - \begin{pmatrix} 0 \\ 0 \\ 2 \end{pmatrix} \right) \right)$$

$$\vec{\nabla}\beta = -\frac{x}{\sigma^2\sqrt{g^{11}}} \begin{pmatrix} 1 \\ 0 \\ 0 \end{pmatrix} - \frac{y}{\sigma^2\sqrt{g^{22}}} \begin{pmatrix} 0 \\ 1 \\ 0 \end{pmatrix} - \frac{k_z}{\sigma^2} \cdot [k_z \cdot z - c \cdot \mathbf{t}] \begin{pmatrix} 0 \\ 0 \\ 1 \end{pmatrix}$$

$$\partial_t\beta = \frac{c}{\sigma^2} \cdot [k_z \cdot z - c \cdot \mathbf{t}]; \quad \vec{C}_{xyz} = C \cdot \begin{pmatrix} y \\ -x \\ 0 \end{pmatrix} \tag{288}$$

and in the case of f:

$$\left\{ \vec{\nabla} \times, \vec{\nabla} \cdot, \partial_t \right\} f_{xyz} \rightarrow e^{\beta} \cdot \left\{ \vec{\nabla} \times \vec{C}_{fxyz} + \vec{\nabla}\beta \times \vec{C}_{fxyz}, \vec{\nabla}\beta \cdot \vec{C}_{fxyz}, \partial_t\beta \cdot \vec{C}_{fxyz} \right\}$$

$$\beta = -\frac{x^2}{2 \cdot \sigma^2 \sqrt{g^{11}}} - \frac{y^2}{2 \cdot \sigma^2 \sqrt{g^{22}}} - \frac{1}{2\sigma^2} \cdot [k_z \cdot z - c \cdot \mathbf{t}]^2$$

$$\vec{\nabla} \times \vec{C}_{fxyz} + \vec{\nabla}\beta \times \vec{C}_{fxyz} = C \cdot \left(\frac{x}{\sigma^2\sqrt{g^{11}}} \begin{pmatrix} 0 \\ 0 \\ x \end{pmatrix} + \frac{y}{\sigma^2\sqrt{g^{22}}} \begin{pmatrix} 0 \\ 0 \\ y \end{pmatrix} \right.$$

$$\left. - \frac{k_z}{\sigma^2} \cdot [k_z \cdot z - c \cdot \mathbf{t}] \left(\begin{pmatrix} x \\ y \\ 0 \end{pmatrix} - \begin{pmatrix} 0 \\ 0 \\ 2 \end{pmatrix} \right) \right)$$

$$\vec{\nabla}\beta = -\frac{x}{\sigma^2\sqrt{g^{11}}} \begin{pmatrix} 1 \\ 0 \\ 0 \end{pmatrix} - \frac{y}{\sigma^2\sqrt{g^{22}}} \begin{pmatrix} 0 \\ 1 \\ 0 \end{pmatrix} - \frac{k_z}{\sigma^2} \cdot [k_z \cdot z - c \cdot \mathbf{t}] \begin{pmatrix} 0 \\ 0 \\ 1 \end{pmatrix}$$

$$\partial_t\beta = \frac{c}{\sigma^2} \cdot [k_z \cdot z - c \cdot \mathbf{t}]; \quad \vec{C}_{fxyz} = C \cdot \begin{pmatrix} y \\ -x \\ 0 \end{pmatrix}. \tag{289}$$

We point out that in the z direction the plane wave does not only stand for the quantum jitter but also gives the principle wave (frequency) for the photon. However, we do not need it for the following discussion.

It looks like that we cannot keep all Maxwell conditions. While solenoidality was easy to be found via a proper vector choice for $\vec{C}_{fxyz}$ and $\vec{C}_{xyz}$, we seem to have problems with the third Maxwell condition the moment we intend to localize our photon.

Thus, either we let fall our third condition and thus, consequently, the vector $\vec{C}_{fxyz}$ would not be orthogonal to $\vec{C}_{xyz}$ anymore. Which means that the f field is not standing rectangular on the F field and the rotation of the one is not equal to the negative time derivative of the other anymore. Where, for now, we do not care which one is standing for the electric and which ones gives the magnetic field. Or alternatively, we keep this orthogonality and sacrifice one of the solenoidality conditions. To sum this up, it looks like that from the current point of view we have the choice of three possible options. These options are:

A. We let fall the solenoidal condition for one of the two fields, either f or F, and force it to fulfill the third Maxwell condition
B. We keep the solenoidal condition and adapt the third Maxwell condition such that we could interpret the additional field term as an yet unknown additional displacement current density
C. The solution is not complete

Option A: Leading to Magnetic Charges

For this example, we set the function vector f to become

$$f_{xyz} = C \cdot \begin{pmatrix} e^{\beta} \cdot x \\ e^{\beta} \cdot y \\ \sqrt{\frac{\pi}{2}} \cdot \frac{e^{\gamma}}{c} \left(2\gamma - 1\right) \cdot erf\left[\frac{k_z \cdot z - c \cdot t}{\sqrt{2} \cdot \sigma z}\right] \end{pmatrix}$$

$$\beta = -\frac{x^2}{2 \cdot \sigma^2 \sqrt{g^{11}}} - \frac{y^2}{2 \cdot \sigma^2 \sqrt{g^{22}}} - \frac{1}{2\sigma^2} \cdot [k_z \cdot z - c \cdot t]^2$$

$$\gamma = -\frac{x^2}{2 \cdot \sigma^2 \sqrt{g^{11}}} - \frac{y^2}{2 \cdot \sigma^2 \sqrt{g^{22}}}. \tag{290}$$

Evaluating the time derivative, we have it fulfilling the third Maxwell equation now with f playing the role of the magnetic field and F presenting

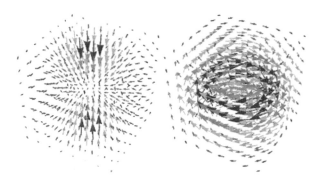

Figure 29 Localized (Gaussian), partially non-solenoidal photon passing along the z axis. Due to the insistence on the third Maxwell condition a magnetic charge results from the f field (left), while the F field still is charge free and has a clearly visible spin (right). Even though the charge is converging to zero extremely quickly "outside" the photon, this "solution" has to be considered unrealistic, simply because it demands the existence of magnetic monopoles.

the electric one. Of course, this is to the cost of the second Maxwell equation and would lead to magnetic charges as shown in Fig. 29.

Option B: Leading to Magnetic Displacement Current Density

For this example, we set the function vectors for F and f as follows:

$$f_{xyz} = C_f \cdot \begin{pmatrix} e^\beta \cdot y \\ -e^\beta \cdot x \\ 0 \end{pmatrix}; \quad F_{xyz} = C_F \cdot \begin{pmatrix} e^\beta \cdot y \\ -e^\beta \cdot x \\ 0 \end{pmatrix}$$

$$\beta = -\frac{x^2}{2 \cdot \sigma^2 \sqrt{g^{11}}} - \frac{y^2}{2 \cdot \sigma^2 \sqrt{g^{22}}} - \frac{1}{2\sigma^2} \cdot [k_z \cdot z - c \cdot t]^2. \quad (291)$$

Now the third Maxwell condition cannot be fulfilled and an additional current density $\vec{J}$ must be introduced in order to connect f and F via the curl and time derivative operator

$$\vec{\nabla} \times f = \vec{J} - \partial_t F. \quad (292)$$

As in the fourth Maxwell equation (see (260)), this current density might be considered a magnetic displacement current density. The left-hand side in Fig. 30 shows the special appearance of this current.

Now the third Maxwell condition cannot be fulfilled and an additional current density $\vec{J}$ must be introduced in order to connect f and F via the curl and time derivative operator.

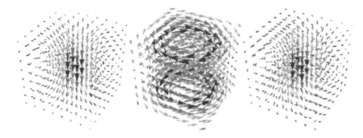

Figure 30 Localized (Gaussian), partially non-solenoidal photon passing along the z axis. Due to the insistence on the soleniodality a magnetic displacement current (right) results the fields. Now both, the f and the F field are showing spin (c.f. right-hand side in Fig. 29). For completeness, we also show the curl (left) and the time derivative fields (middle). Still, this "solution" has to be considered unrealistic or at least incomplete, even though the displacement current drops extremely fast to zero (c.f. also Fig. 31).

Option C: Finding the Correct Metric, a Yet Unsolved Problem

It is clear from the considerations above, that a metric with only constants cannot be the correct metric for the photon, because it does not satisfy all necessary boundary conditions. We saw, however, that our approach of a constant (non-coordinate dependent) metric with shear provides plane wave solutions, which could be superposed such that localized objects can be obtained. One might interpret these objects as photons. Summing up what we have gained so far, we can write

$$\vec{t} = \mathbf{x_0} = \begin{pmatrix} G_0 \\ 0 \\ 0 \\ 0 \end{pmatrix} + \begin{pmatrix} E_0 \\ 0 \\ 0 \\ E_{30} \end{pmatrix} f_0\,(t, x, y, z)\,;$$

$$\vec{x} = \mathbf{x_1} = \begin{pmatrix} 0 \\ G_1 \\ 0 \\ 0 \end{pmatrix} + \begin{pmatrix} 0 \\ E_1 \\ 0 \\ 0 \end{pmatrix} f_1\,(t, x, y, z)\,;$$

$$\vec{y} = \mathbf{x_2} = \begin{pmatrix} 0 \\ 0 \\ G_2 \\ 0 \end{pmatrix} + \begin{pmatrix} 0 \\ 0 \\ E_2 \\ 0 \end{pmatrix} f_2\,(t, x, y, z)\,;$$

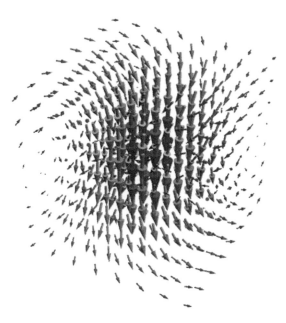

Figure 31 Magnetic displacement current field resulting from our incomplete solution.

$$\vec{z} = \mathbf{x}_3 = \begin{pmatrix} 0 \\ 0 \\ 0 \\ G_3 \end{pmatrix} + \begin{pmatrix} E_{30} \\ 0 \\ 0 \\ E_3 \end{pmatrix} f_3\,(t, x, y, z) \tag{293}$$

where we assume to have $f_0 = f_{xyz}$ and $f_3 = F_{xyz}$ and we will see what we can do for the functions f_1 and f_2 later on. So far we have (with the option to additional $i*x$-$i*y$ exponents as shown in (279))

$$F_{xyz} = c_x \cdot c_y \cdot \vec{\mathbf{C}}_F \cdot e^{-\frac{x^2}{2\cdot\sigma^2\sqrt{g^{11}}} - \frac{y^2}{2\cdot\sigma^2\sqrt{g^{22}}} - \frac{1}{2\sigma^2}\cdot[k_z\cdot z - c\cdot t]^2 + i\cdot\alpha_{F0}\cdot[k_z\cdot z - c\cdot t]}$$

$$= C \cdot \begin{pmatrix} y \\ -x \\ 0 \end{pmatrix} \cdot e^{-\frac{x^2}{2\cdot\sigma^2\sqrt{g^{11}}} - \frac{y^2}{2\cdot\sigma^2\sqrt{g^{22}}} - \frac{1}{2\sigma^2}\cdot[k_z\cdot z - c\cdot t]^2 + i\cdot\alpha_{F0}\cdot[k_z\cdot z - c\cdot t]}, \tag{294}$$

$$f_{xyz} = c_x \cdot c_y \cdot \vec{\mathbf{C}}_f \cdot e^{-\frac{x^2}{2\cdot\sigma^2\sqrt{g^{11}}} - \frac{y^2}{2\cdot\sigma^2\sqrt{g^{22}}} - \frac{1}{2\sigma^2}\cdot[k_z\cdot z - c\cdot t]^2 + i\cdot\alpha_{f0}\cdot[k_z\cdot z - c\cdot t]}$$

$$= C \cdot \begin{pmatrix} y \\ -x \\ 0 \end{pmatrix} \cdot e^{-\frac{x^2}{2\cdot\sigma^2\sqrt{g^{11}}} - \frac{y^2}{2\cdot\sigma^2\sqrt{g^{22}}} - \frac{1}{2\sigma^2}\cdot[k_z\cdot z - c\cdot t]^2 + i\cdot\alpha_{f0}\cdot[k_z\cdot z - c\cdot t]}. \tag{295}$$

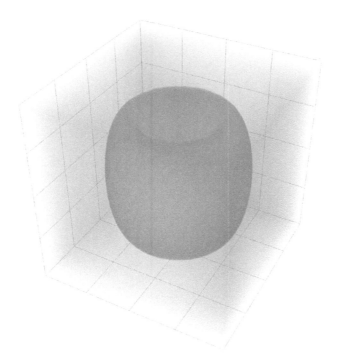

Figure 32 Energy density plot of a photon with our solenoidal approach traveling along the z axis.

With the two functions f and F being—somehow—connected with the electric and magnetic field of the photon, we should be able to find the energy density of this particle according to our approach. This energy density can be evaluated as being proportional to $f^2 + F^2$ and is shown in Fig. 32.

As we have seen that the problems we faced with respect to the Maxwell equations are arising from the localization of the photon in the direction perpendicular to the propagation we now seek for a general approach with an x-y dependency fulfilling the following conditions:

$$\begin{pmatrix} \partial_x \\ \partial_y \\ 0 \end{pmatrix} \cdot \begin{pmatrix} f_x\,(x,y) \\ f_y\,(x,y) \\ 0 \end{pmatrix} = 0; \quad \partial_x f_y\,(x,y) - \partial_y f_x\,(x,y) = 0. \tag{296}$$

With the expression $\Phi\,[\ldots]$ denoting function of $[\ldots]$ we give the general solution as follows:

$$f_x\,(x,y) = i \cdot C_1 \cdot \Phi\,[ix + y] - i \cdot C_2 \cdot \Phi\,[-ix + y]$$
$$f_y\,(x,y) = C_1 \cdot \Phi\,[ix + y] + C_2 \cdot \Phi\,[-ix + y]. \tag{297}$$

Now we adapt our two functions f and F as follows:

$$F_{xyz} = C \cdot \begin{pmatrix} f_{Fx}(x, y) \\ f_{Fy}(x, y) \\ 0 \end{pmatrix} \cdot e^{-\frac{1}{2\sigma^2} \cdot [k_z \cdot z - c \cdot t]^2 + i \cdot \alpha_{F0} \cdot [k_z \cdot z - c \cdot t]}$$

$$\equiv C \cdot \begin{pmatrix} f_{Fx}(x, y) \\ f_{Fy}(x, y) \\ 0 \end{pmatrix} \cdot g_F(z, t)$$

$$f_{xyz} = C \cdot \begin{pmatrix} f_{fx}(x, y) \\ f_{fy}(x, y) \\ 0 \end{pmatrix} \cdot e^{-\frac{1}{2\sigma^2} \cdot [k_z \cdot z - c \cdot t]^2 + i \cdot \alpha_{f0} \cdot [k_z \cdot z - c \cdot t]}$$

$$\equiv C \cdot \begin{pmatrix} f_{fx}(x, y) \\ f_{fy}(x, y) \\ 0 \end{pmatrix} \cdot g_f(z, t)$$

$$f_{Fx}(x, y) = i \cdot C_{1F} \cdot \Phi[ix + y] - i \cdot C_{2F} \cdot \Phi[-ix + y]$$
$$f_{Fy}(x, y) = C_{1F} \cdot \Phi[ix + y] + C_{2F} \cdot \Phi[-ix + y]$$
$$f_{fx}(x, y) = i \cdot C_{1f} \cdot \Phi[ix + y] - i \cdot C_{2f} \cdot \Phi[-ix + y]$$
$$f_{fy}(x, y) = C_{1f} \cdot \Phi[ix + y] + C_{2f} \cdot \Phi[-ix + y]. \qquad (298)$$

We remind (c.f. (255) and (256)) that the g functions in general stand for the more general wave-expression:

$$g_{F,f} \to \text{Function}\left[\begin{pmatrix} 0 \\ 0 \\ k_{Fz, fz} \end{pmatrix} \cdot \begin{pmatrix} 0 \\ 0 \\ z \end{pmatrix} - c \cdot t \right]. \qquad (299)$$

Both functions F and f are solenoidal, their curl gives non-zero expressions only for the two first vector components and the time derivative can now be combined with the curl as demanded by Maxwell's third law, because we have

$$\vec{\nabla} \times F_{xyz} = C \cdot \begin{pmatrix} -f_{Fy}(x, y) \\ f_{Fx}(x, y) \\ 0 \end{pmatrix} \cdot g_F^{(1,0)}(z, t);$$

$$\vec{\nabla} \times f_{xyz} = C \cdot \begin{pmatrix} -f_{fy}(x, y) \\ f_{fx}(x, y) \\ 0 \end{pmatrix} \cdot g_f^{(1,0)}(z, t)$$

$$\partial_t F_{xyz} = C \cdot \begin{pmatrix} f_{Fx}(x, y) \\ f_{Fy}(x, y) \\ 0 \end{pmatrix} \cdot g_F^{(0,1)}(z, t);$$

$$\partial_t f_{xyz} = C \cdot \begin{pmatrix} f_{fx}(x, y) \\ f_{fy}(x, y) \\ 0 \end{pmatrix} \cdot g_f^{(0,1)}(z, t). \tag{300}$$

We see that the third Maxwell condition, which could either be given as $\vec{\nabla} \times f = -\partial_t F$ or as $\vec{\nabla} \times F = -\partial_t f$, forces us to set

$$\begin{pmatrix} -f_{fy}(x, y) \\ f_{fx}(x, y) \\ 0 \end{pmatrix} \cdot g_f^{(1,0)}(z, t) = - \begin{pmatrix} f_{Fx}(x, y) \\ f_{Fy}(x, y) \\ 0 \end{pmatrix} \cdot g_F^{(0,1)}(z, t)$$

or $\tag{301}$

$$\begin{pmatrix} -f_{Fy}(x, y) \\ f_{Fx}(x, y) \\ 0 \end{pmatrix} \cdot g_F^{(1,0)}(z, t) = - \begin{pmatrix} f_{fx}(x, y) \\ f_{fy}(x, y) \\ 0 \end{pmatrix} \cdot g_f^{(0,1)}(z, t).$$

As chosen previously (c.f. (264)) we use the first choice which gives us the electric and magnetic field as (please note: f and F are vector functions)

$$\vec{E} = f; \quad \vec{B} = F \tag{302}$$

and we obtain the following equation to solve

$$\begin{pmatrix} -C_{1f} \cdot \Phi[ix + y] - C_{2f} \cdot \Phi[-ix + y] \\ i \cdot C_{1f} \cdot \Phi[ix + y] - i \cdot C_{2f} \cdot \Phi[-ix + y] \\ 0 \end{pmatrix} \cdot g_f'(z, t) \cdot k_{fz}$$

$$= \begin{pmatrix} i \cdot C_{1F} \cdot \Phi[ix + y] - i \cdot C_{2F} \cdot \Phi[-ix + y] \\ C_{1F} \cdot \Phi[ix + y] + C_{2F} \cdot \Phi[-ix + y] \\ 0 \end{pmatrix} \cdot g_F'(z, t) \cdot c \tag{303}$$

$$g_{f,F}'(z, t) = \frac{\partial g_{f,F}([k_{fz,Fz} \cdot z - c \cdot t])}{\partial[k_{fz,Fz} \cdot z - c \cdot t]}$$

We simplify the expression above by setting both g functions to be equal (meaning that we also have $k_{Fz} = k_{fz} = k_z$), which results in

$$C_{1F} = \frac{i \cdot C_{1f} \cdot k_z}{c}; \quad C_{2F} = -\frac{i \cdot C_{2f} \cdot k_z}{c}. \tag{304}$$

Thus, we can finally write down our 3D wave function for a photonic object (for brevity we suck the constant C into the constants $C1_f$ and $C2_f$,

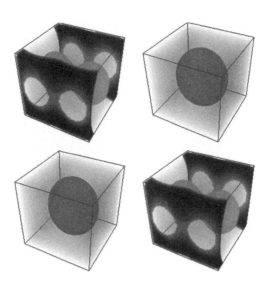

Figure 33 Electric (above) and magnetic field density (below) of a photon with our solenoidal approach traveling along the z axis. On the left-hand side, we have the x component and on the right-hand side the y component.

c.f. Eq. (298)):

$$F_{xyz} = \tfrac{k_z}{c} \cdot \begin{pmatrix} -C_{1f} \cdot \Phi\,[ix+y] - C_{2f} \cdot \Phi\,[-ix+y] \\ i \cdot C_{1f} \cdot \Phi\,[ix+y] - i \cdot C_{2f} \cdot \Phi\,[-ix+y] \\ 0 \end{pmatrix} \cdot g\,([k_z \cdot z - c \cdot t])$$

$$f_{xyz} = \begin{pmatrix} i \cdot C_{1f} \cdot \Phi\,[ix+y] - i \cdot C_{2f} \cdot \Phi\,[-ix+y] \\ C_{1f} \cdot \Phi\,[ix+y] + C_{2f} \cdot \Phi\,[-ix+y] \\ 0 \end{pmatrix} \cdot g\,([k_z \cdot z - c \cdot t])$$

$$(305)$$

Now assuming g functions of the type given in (298) and similar forms for the $\Phi\,(x, y)$ forms only that we demand the absolute values for the squared term, because of the imaginary $i*x$, we can derive some impressive illustrations for the photon (Figs. 33 to 35).

Suspicion about Connections to Compactified Coordinates

As we have seen, the Maxwell conditions 1 to 3 could only be satisfied by the introduction of the functions $\Phi\,(x, y)$, which are nothing else but

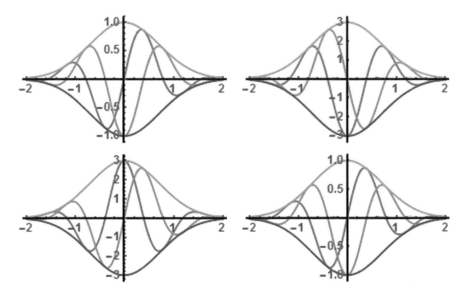

Figure 34 Presentation of x and y part of functions f (upper pair left and right, respectively) and F (lower pair left and right, respectively) with blue giving the real, yellow ocher the imaginary part and red and green showing the upper and lower envelope.

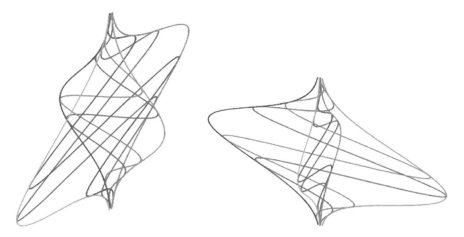

Figure 35 Presentation of x (left) and y part (right) of functions f and F as parametric 3D plot with f bending in x direction and F bending in y direction (blue & violet = real, yellow ochre & brown = imaginary and red, green & yellow, bright blue showing the upper and lower envelope).

wave like connections between a real x or y axis with the corresponding imaginary counterpart $i*y$ or $i*x$, respectively (c.f. (305)). We take this as hint for hidden (compactified) coordinates and interpret (only as a possibility) the $i*x$ in the function $\Phi(x, y) = \Phi(y - i \cdot x)$ as the radius of the x axis. We simply assume the y axis to be more like a hose or tube of a certain radius rather than a line of zero thickness. We further assume that this radius can vary and that it oscillates in the case of a passing photon. This hidden dimension of the y coordinate (imagined as radius) reveals itself within our equations in form of the imaginary $i*x$. The same can be said about the x coordinate where now $i*y$ takes on the hidden dimension. Real and imaginary part of our functions $\Phi(x, y) = \Phi(y - i \cdot x)$ are then just longitudinal and radius waves of the tube-like coordinates x and y. Figures 36 to 38 illustrate this situation in various forms. It is clear that according to the substructural hypothesis necessary also to explain other aspects of space-time, these illustrations can only be considered approximations. We leave it to the reader to, perhaps, find even better illustrations of the photon.

Figure 36 Presentation of functions f and F as parametric plot with the f bending in x direction and the F beding in y direction. This time the imaginary part defines the thickness of the tubes. The y part only interchanges f and F and therefore is not drawn here. In addition the real part controls the color of the tube.

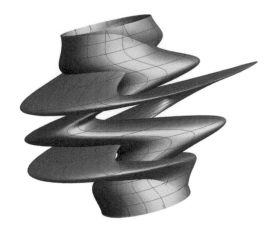

Figure 37 Presentation of f as parametric plot with the real part (blue) and the imaginary part (yellow ochre). This time the z axis is shown as two tubes being bent and changed in radius according to the strength of the field.

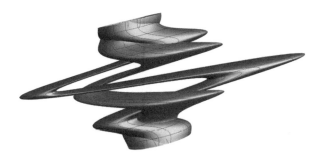

Figure 38 Presentation of F as parametric plot with the real part (blue) and the imaginary part (yellow ochre). This time the z axis is shown as two tubes being bent and changed in radius according to the strength of the field.

The Alternative Interpretation Using Real and Imaginary Part

Again, we start with our general solutions f and F in the following form:

$$F_{xyz} = \begin{pmatrix} f_{Fx}\,(x, y) \\ f_{Fy}\,(x, y) \\ 0 \end{pmatrix} \cdot g_F\,(z, t)\,; \quad f_{xyz} = \begin{pmatrix} f_{fx}\,(x, y) \\ f_{fy}\,(x, y) \\ 0 \end{pmatrix} \cdot g_f\,(z, t)$$

$$f_{Fx}\,(x, y) = \mathrm{i} \cdot C_{1F} \cdot \Phi\,[\mathrm{i}x + y] - \mathrm{i} \cdot C_{2F} \cdot \Phi\,[-\mathrm{i}x + y]$$

$$f_{Fy}(x, y) = C_{1F} \cdot \Phi[ix + y] + C_{2F} \cdot \Phi[-ix + y]$$

$$f_{fx}(x, y) = i \cdot C_{1f} \cdot \Phi[ix + y] - i \cdot C_{2f} \cdot \Phi[-ix + y]$$

$$f_{fy}(x, y) = C_{1f} \cdot \Phi[ix + y] + C_{2f} \cdot \Phi[-ix + y]. \tag{306}$$

As said before, both functions F and f are solenoidal, their curl gives non-zero expressions only for the two first vector components and the time derivative can now be combined with the curl as demanded by Maxwell's third law, because we have

$$\vec{\nabla} \times F_{xyz} = C \cdot \begin{pmatrix} -f_{Fy}(x, y) \\ f_{Fx}(x, y) \\ 0 \end{pmatrix} \cdot g_F^{(1,0)}(z, t);$$

$$\vec{\nabla} \times f_{xyz} = C \cdot \begin{pmatrix} -f_{fy}(x, y) \\ f_{fx}(x, y) \\ 0 \end{pmatrix} \cdot g_f^{(1,0)}(z, t)$$

$$\partial_t F_{xyz} = C \cdot \begin{pmatrix} f_{Fx}(x, y) \\ f_{Fy}(x, y) \\ 0 \end{pmatrix} \cdot g_F^{(0,1)}(z, t);$$

$$\partial_t f_{xyz} = C \cdot \begin{pmatrix} f_{fx}(x, y) \\ f_{fy}(x, y) \\ 0 \end{pmatrix} \cdot g_f^{(0,1)}(z, t). \tag{307}$$

Again we have sucked the constant C into the new constants $C1_{f,F}$ and $C2_{f,F}$ of our functions $f_{Fx, Fy, fx, fy}$. In contrast to our approach in section "The other way to fulfill the Maxwell equations with plane waves" this time we do not set the third Maxwell condition as $\vec{\nabla} \times f = -\partial_t F$ or as $\vec{\nabla} \times F = -\partial_t f$ but demand instead

$$\vec{E} = \text{Re}[f + F]; \quad \vec{B} = C \cdot \text{Im}[f - F] \tag{308}$$

with

$$\vec{\nabla} \times \text{Re}[f + F] = -C \cdot \partial_t \text{Im}[f - F]. \tag{309}$$

Here we introduced a new constant C, which is required in order to provide a proper unit connection between the wave number k_z and the speed of light in vacuum c. It has nothing to do with our previous constant C for f and F.

We obtain the following equation to solve

$$\text{Re}\left[\begin{pmatrix} -\left(C_{1f}+C_{1F}\right)\cdot\Phi\left[ix+y\right]-\left(C_{2f}+C_{2F}\right)\cdot\Phi\left[-ix+y\right] \\ i\cdot\left(C_{1f}+C_{1F}\right)\cdot\Phi\left[ix+y\right]-i\cdot\left(C_{2f}+C_{2F}\right)\cdot\Phi\left[-ix+y\right] \\ 0 \end{pmatrix}\cdot g'_f(z,t)\right]\cdot\frac{k_{fz}}{C}$$

$$=\text{Im}\left[\begin{pmatrix} i\cdot\left(C_{1f}-C_{1F}\right)\cdot\Phi\left[ix+y\right]-i\cdot\left(C_{2f}-C_{2F}\right)\cdot\Phi\left[-ix+y\right] \\ \left(C_{1f}-C_{1F}\right)\cdot\Phi\left[ix+y\right]+\left(C_{2f}-C_{2F}\right)\cdot\Phi\left[-ix+y\right] \\ 0 \end{pmatrix}\cdot g'_F(z,t)\right]\cdot c$$

$$g'_{f,F}(z,t)=\frac{\partial g_{f,F}\left(\left[k_{fz,Fz}\cdot z-c\cdot t\right]\right)}{\partial\left[k_{fz,Fz}\cdot z-c\cdot t\right]};\quad k_{Fz}=k_{fz}=k_z. \tag{310}$$

We intend to simplify the evaluation by separating the equations as follows:

$$\text{Re}\left[\begin{pmatrix} -\left(C_{1f}+C_{1F}\right)\cdot\Phi\left[ix+y\right] \\ i\cdot\left(C_{1f}+C_{1F}\right)\cdot\Phi\left[ix+y\right] \\ 0 \end{pmatrix}\right]\cdot\frac{k_z}{C}$$

$$=\text{Im}\left[\begin{pmatrix} i\cdot\left(C_{1f}-C_{1F}\right)\cdot\Phi\left[ix+y\right] \\ \left(C_{1f}-C_{1F}\right)\cdot\Phi\left[ix+y\right] \\ 0 \end{pmatrix}\right]\cdot c$$

$$\text{Re}\left[\begin{pmatrix} -\left(C_{2f}+C_{2F}\right)\cdot\Phi\left[-ix+y\right] \\ -i\cdot\left(C_{2f}+C_{2F}\right)\cdot\Phi\left[-ix+y\right] \\ 0 \end{pmatrix}\right]\cdot\frac{k_z}{C}$$

$$=\text{Im}\left[\begin{pmatrix} -i\cdot\left(C_{2f}-C_{2F}\right)\cdot\Phi\left[-ix+y\right] \\ \left(C_{2f}-C_{2F}\right)\cdot\Phi\left[-ix+y\right] \\ 0 \end{pmatrix}\right]\cdot c, \tag{311}$$

which gives us the simple equations

$$\text{Re}\begin{bmatrix} -\left(C_{1f}+C_{1F}\right) \\ i\cdot\left(C_{1f}+C_{1F}\right) \\ 0 \end{bmatrix}\cdot\frac{k_z}{C}=\text{Im}\begin{bmatrix} i\cdot\left(C_{1f}-C_{1F}\right) \\ \left(C_{1f}-C_{1F}\right) \\ 0 \end{bmatrix}\cdot c$$

$$\text{Re}\begin{bmatrix} -\left(C_{2f}+C_{2F}\right) \\ -i\cdot\left(C_{2f}+C_{2F}\right) \\ 0 \end{bmatrix}\cdot\frac{k_z}{C}=\text{Im}\begin{bmatrix} -i\cdot\left(C_{2f}-C_{2F}\right) \\ \left(C_{2f}-C_{2F}\right) \\ 0 \end{bmatrix}\cdot c, \tag{312}$$

with the following subsequent solution:

$$\text{Re}\left[C_{1F}\right]=\text{Re}\left[C_{1f}\right]\cdot\frac{c\cdot C+k_z}{c\cdot C-k_z};\quad \text{Im}\left[C_{1F}\right]=\text{Im}\left[C_{1f}\right]\cdot\frac{c\cdot C+k_z}{c\cdot C-k_z}$$

$$\text{Re}\left[C_{2F}\right]=\text{Re}\left[C_{2f}\right]\cdot\frac{c\cdot C-k_z}{c\cdot C+k_z};\quad \text{Im}\left[C_{2F}\right]=\text{Im}\left[C_{2f}\right]\cdot\frac{c\cdot C-k_z}{c\cdot C+k_z}. \tag{313}$$

Further Illustrations and a Few Words about the Absence of Magnetic Monopoles in Our Observable Universe

With this constellation we could interpret one of the fields—in this case the magnetic one—as the one being solely connected with the hidden—compactified—dimensions i^*x and i^*y. They appear in addition to the known spatial dimensions x, y and z. As the photon can travel in all possible spatial directions and our photonic solutions (306) with its field interpretations (308) should therefore also be written for x, y and any other propagation direction in the same way, such a winded-up dimension must also be present for z, of course. Thus, we conclude that there must be at least one additional set of i^*x, i^*y and i^*z as compactified coordinates. The substructure suggested by the author a while ago fits well to this assumption (f.c. [17, 23]). A Matryoshka-like setup of space with 2-spheres, named "Friedmanns" in [1, 2], building the grainy space and being built themselves out of much smaller 2-spheres, mainly forming the surfaces (membranes) of the bigger spheres and so on, on the various levels of scales, could also explain the absence of magnetic monopoles (see further below). As we can see from our setting in (308) and the subsequent derivations, the total localized photonic field contains a classical wave function propagating with the direction of light (which in our case was z) and a complex $i^*x \pm y$-term for the axes perpendicular to z. These terms could also be interpreted as waves, namely as waves between the ordinary and the compactified coordinates. With our setting (308) we then declare the compactified—imaginary—part to form the magnetic field with waves respectively oscillations restricted to these very compactified coordinates. In reality, however, we do not have to deal with completely new ominous dimensions, but simply with additional degrees of freedom arising from the substructure. So, the spheres forming our space could simply be deformed in a variety of ways allowing for some additional properties not being present in the perfectly smooth and continuous three-dimensional space. As the many spheres are difficult to draw we could resort to the illustrative approximation of tubes winding themselves along the direction of propagation of a photon and showing certain deformations according to the field strength given throughout our solutions (306) with the setting (308) and the coefficients (313). Thus, as said before, we can illustrate a localized photon by using an ensemble of

Figure 39 Presentation of electric (yellow) and magnetic (blue) field. The z axis is shown as two tubes being bent according to the strength of the fields.

2-spheres or more easy tubes (understood as summed up spheres) and deform them in accordance to the electric and magnetic field, respectively the real and imaginary values of our solutions. At first we consider the electric and magnetic part separately (two tubes, see Fig. 39) and discover that there is not much difference to our illustrations before (c.f. Figs. 37 and 38).

Now, however, with our clear assignment of the two fields to

(a) deformation (bending) of the tube's axes in total in case of the electric and
(b) deformation of the tube's wall, changing its diameter, in the case of the magnetic field

we can put all field information on just one tube (Fig. 40).

We recognize the double deformation of the "coordinate-tube" as x-y deformation of its symmetry axis (electric field) and its radius shape deformation as result of the magnetic field along the z axis (Fig. 40) for the passing photon. In Fig. 41 we have also coded the local energy density given as $E^2 + B^2$ in the color of the distorted tube.

With the magnetic field being "hidden" within the compactified coordinates or degrees of freedom, the magnetic field effects can only indirectly be detected on the scale of our grainy space. Thus, magnetic monopoles might well exist within the compactified coordinates residing there as stable oscillations or standing waves, but they cannot be observed as elementary particles on our level, because any deformation of the hidden coordinates is only detectable via its effects on the "ordinary" coordinates.

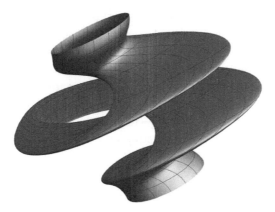

Figure 40 Presentation of electric (axis deformation) and magnetic (radius distortion) field. The *z* axis is shown as one tube being bent and distorted in radius according to the strength of the fields.

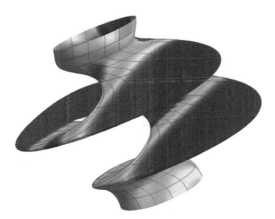

Figure 41 Presentation of electric (axis deformation) and magnetic (radius distortion) field. The *z* axis is shown as one tube being bent and distorted in radius according to the strength of the fields. In addition the color is coding the energy density of the passing photon.

Within our illustration of the electric and magnetic field for the photon, the magnetic monopole would become visible at our scale throughout its indirect distortion of the space but not its original deformation of the tube's cylindrical shape. Assuming now a magnetic monopole particle to be some kind of permanent such shape deformation, we will not see it, but only its electrical counterpart from the resulting "ordinary spatial deformation." Only within oscillations and other time-dependent effects as with the photon, we detect the mixed or two-fold character of the electromagnetic field, because of the time shift between the two components. As static or quasi-static effect it cannot exist (for us), because we only see deformations on our observable ("direct") coordinates and there is no magnetic field, which is to say no magnetic—compactified coordinate—distortion. Only the other field part, the electric one, which is directly acting on our spatial coordinates can be observed directly.

Things are probably vice versa within the scale level underneath ours. There, one will not find electric monopoles, but magnetic ones instead.

The Total Spatial Displacement for the Photon

Now that we have completed the solution for the spatial displacement and assigned all results to our classical electric and magnetic fields, we should apply equation (293) again and see how the photon does act on our four dimensional space-time. With the necessary settings (275) and (276) we obtain the following interesting solution for our base-vector components E_i from equation (272):

$$\vec{t} = \mathbf{x_0} = \begin{pmatrix} G_0 \\ 0 \\ 0 \\ 0 \end{pmatrix} + \begin{pmatrix} E_0 \\ 0 \\ 0 \\ E_{30} \end{pmatrix} f_0\,(t, x, y, z)\,;$$

$$\vec{x} = \mathbf{x_1} = \begin{pmatrix} 0 \\ G_1 \\ 0 \\ 0 \end{pmatrix} + \begin{pmatrix} 0 \\ E_1 \\ 0 \\ 0 \end{pmatrix} f_1\,(t, x, y, z)\,;$$

$$\vec{y} = \mathbf{x_2} = \begin{pmatrix} 0 \\ 0 \\ G_2 \\ 0 \end{pmatrix} + \begin{pmatrix} 0 \\ 0 \\ E_2 \\ 0 \end{pmatrix} f_2\,(t, x, y, z)\,;$$

$$\vec{z} = \mathbf{x_3} = \begin{pmatrix} 0 \\ 0 \\ 0 \\ G_3 \end{pmatrix} + \begin{pmatrix} E_{30} \\ 0 \\ 0 \\ E_3 \end{pmatrix} f_3\,(t, x, y, z) \tag{314}$$

with

$$
\begin{pmatrix} G_0 \\ 0 \\ 0 \\ 0 \end{pmatrix} + \begin{pmatrix} E_0 \\ 0 \\ 0 \\ E_{30} \end{pmatrix} = \begin{pmatrix} \pm\sqrt{\frac{g^{30}}{c\cdot(1+c^2)}} \\ 0 \\ 0 \\ \pm\sqrt{\frac{c\cdot g^{30}}{1+c^2}} \end{pmatrix} \quad ; \qquad \begin{pmatrix} 0 \\ G_1 \\ 0 \\ 0 \end{pmatrix} + \begin{pmatrix} 0 \\ E_1 \\ 0 \\ 0 \end{pmatrix} = \begin{pmatrix} 0 \\ \pm\sqrt{g^{11}} \\ 0 \\ 0 \end{pmatrix} \quad ;
$$

$$
\begin{pmatrix} 0 \\ 0 \\ G_2 \\ 0 \end{pmatrix} + \begin{pmatrix} 0 \\ 0 \\ E_2 \\ 0 \end{pmatrix} = \begin{pmatrix} 0 \\ 0 \\ \pm\sqrt{g^{11}} \\ 0 \end{pmatrix} \quad ; \qquad \begin{pmatrix} 0 \\ 0 \\ 0 \\ G_3 \end{pmatrix} + \begin{pmatrix} E_{30} \\ 0 \\ 0 \\ E_3 \end{pmatrix} = \begin{pmatrix} \pm\sqrt{\frac{c\cdot g^{30}}{1+c^2}} \\ 0 \\ 0 \\ \pm c\cdot\sqrt{\frac{c\cdot g^{30}}{1+c^2}} \end{pmatrix}
$$

$$\tag{315}$$

where the sign of first and fourth base vector are connected (that is why the cursive "$\pm$" signs there), giving 8 possible tuples of base vectors. Most interestingly the necessary settings (275) and (276) are leading to $E_{30} = c*E_0 = E_3/c$. Thus, nature seems to prefer a fine-tuning of the metric coefficients g^{00} and g^{33} such that even though there is a metric shear g^{30} in time and space, the resulting base vectors do show a perfect symmetry connected via the speed of light in the form

$$
\mathbf{e}_0 = \begin{pmatrix} E_0 \\ 0 \\ 0 \\ E_{30} \end{pmatrix} = \frac{\mathbf{e}_3}{c} = \frac{1}{c}\begin{pmatrix} E_{30} \\ 0 \\ 0 \\ E_3 \end{pmatrix}.
$$

$$\tag{316}$$

In a unit system with $c = 1$ they'd even be equal and could be seen as just one independent coordinate.

How can we interpret this finding? It seems that when there is shear between time and a spatial coordinate, nature forms a state (in this case a photon) dissolving the shear by an apparent reduction of the dimension of space-time. The formerly four independent dimensions are becoming three. Thus, stable states like particles apparently require the existence of a base of the form (315) or (316), which is to say a dimensional reduction via the connection of a spatial dimension and time. Illustratively put into a three-dimensional example one could say: What to us looks like a speedy particle in reality is nothing but a moving cuboid squashed to a plane (from the limited perspective of our spatial scale level). What has been time and z axis before became a new time-axis for the remaining 2D state named photon (thereby not counting the hidden parameters respectively the compactified coordinates).

Conclusions to the Photon

In this work, the author has shown how a most simple, but Einstein-field-compatible metric with constant diagonal and shear components gives solutions of wave character and the necessary properties required for the photon.

The solutions of the subsequent quantization of the metric lead to the typical plane waves known from the classical Maxwell theory.

While these plane wave solutions allow for all Maxwell conditions to be fulfilled, one obtains difficulties the moment one intends to make single photons localized objects in space-time.

Demanding solenoidal properties, which is to say free-off charge character for such single photons, immediately gives these photons spin 1, which agrees well with the standard theories. Thus, two of the Maxwell conditions, namely those demanding zero divergence for electric and magnetic part of the photon, can be satisfied with the introductions of spin.

If insisting on purely real solutions, the third Maxwell condition, however, which connects curl of electric field and time derivative of the magnetic one, requires the introduction of a magnetic displacement current. Thus, it was concluded that the assumption of a purely real solution is too simple for a complete description of the photon.

The introduction of complex solutions finally solves all Maxwell conditions but gives rise to the conclusion of hidden, which is to say "compactified" dimensions.

It was shown how these additional dimensions could be interpreted as hose- or tube-like classical space dimensions forming surfaces instead of one-dimensional lines. Further, it was concluded that our electric and magnetic fields are just oscillations on these surfaces.

In addition, it was found that hidden or compactified dimensions are necessary to fulfill all Maxwell conditions and that the assignment of the magnetic field to these coordinates explains the absence of magnetic monopoles in our observable universe.

Evaluating the new base vectors of the photon-deformed space-time one discovers that the photon reduces the total number of dimension of space-time from four to three (not counting the compactified dimensions).

Chapter 10

How the Quantum Theory Already Resides in the Einstein–Hilbert Action

Theory: The Discarded Term

In order to properly understand why the present author is convinced that Einstein and Hilbert "already had, but did not see it," we resort to small repetition of the mathematical basics and some historical facts of the so called Einstein–Hilbert action [4]. This extremal task, given as

$$\delta_g W = 0 = \delta_g \int_V d^n x \left(\sqrt{-g} \left[R - 2\kappa L_M \right] \right), \tag{317}$$

was submitted by Hilbert in 1915 about 5 days before Einstein submitted his basics of the general theory of relativity [3]. One has to admit, however, that Hilbert's work based on Einstein's previous publications on the matter. Albert Einstein [3] has shown that a space of a given metric $g^{\alpha\beta}$ must curve and that the curvature described by the Ricci tensor $R^{\alpha\beta}$ must satisfy certain conditions. The result for the metric tensor $g^{\alpha\beta}$ of the curved space as given by Einstein can be given as follows:

$$R^{\alpha\beta} - \frac{1}{2} R g^{\alpha\beta} + \Lambda g^{\alpha\beta} = -\kappa T^{\alpha\beta}. \tag{318}$$

Here we have $R^{\alpha\beta}$, $T^{\alpha\beta}$ the Ricci and the energy momentum tensor, respectively, while the parameters Λ and κ are constants (usually called

The Theory of Everything: Quantum and Relativity is Everywhere – A Fermat Universe
Norbert Schwarzer
Copyright © 2020 Jenny Stanford Publishing Pte. Ltd.
ISBN 978-981-4774-47-5 (Hardcover), 978-1-315-09975-0 (eBook)
www.jennystanford.com

cosmological and coupling constant, respectively). These are the well-known Einstein field equations in n dimensions with the indices α and β running from 1 to n. The theory behind is called "general theory of relativity." Hilbert had shown in [4] that this set of equations could be derived from a minimum principle using the so-called Einstein–Hilbert action. This classical Einstein–Hilbert action is given in (317), with g denoting the determinant of the metric tensor, W and V giving the action and the volume of the n-dimensional space, respectively, R denoting the scalar curvature or Ricci scalar, and the last term $2\kappa L_M$ describing the Lagrange density of matter. This action avoids higher orders of curvature respectively higher orders for the Ricci scalar and/or the metric (within the Lagrangian approach as used by Hilbert [4]).

The author assumes that the higher orders are excluded by "the universe," because they would rule out integer solutions according to the Fermat principle as shown in (c.f. section "About Fermat's last theorem" of this book).

It is important to note that neither Einstein nor Hilbert were able to give a proper explanation for the appearance of matter and so they had to postulate it, which they did by the introduction of the energy momentum tensor, respectively, its Lagrange equivalent in the variation (317). This means, however, that the matter term $T^{\alpha\beta}$ cannot be derived directly out of the general theory of relativity, but has to be postulated and fitted in accordance to the problem in question.

In [27] a procedure was suggested by the present author, possibly explaining the occurrence of matter due to a subspace level rather than postulating it.

Taking this suggestion of the subspace level, however, we might like to reconsider some of the basic assumptions in Hilbert's old derivation of the Einstein field equations [4]. Therefore, we will need the following variations δ_g with respect to the metric tensor $g^{\alpha\beta}$:

$$\delta_g \left(g^{\alpha\beta} \right) = -\frac{1}{2} \left(g^{\alpha\kappa} g^{\beta\lambda} + g^{\beta\kappa} g^{\alpha\lambda} \right) \delta_g g_{\kappa\lambda}$$

$$g^{\alpha\beta}_{,\sigma} = \partial_\sigma g^{\alpha\beta} = \frac{\partial g^{\alpha\beta}}{\partial x^\sigma} = -\frac{1}{2} \left(g^{\alpha\kappa} g^{\beta\lambda} + g^{\beta\kappa} g^{\alpha\lambda} \right) \partial_\sigma g_{\kappa\lambda}$$

$$\Gamma^\gamma_{\alpha\beta} = \frac{g^{\gamma\sigma}}{2} \left(g_{\sigma\alpha,\beta} + g_{\sigma\beta,\alpha} - g_{\alpha\beta,\sigma} \right) \equiv \frac{g^{\gamma\sigma}}{2} (\ldots)_{\alpha\beta\sigma}$$

$$\Rightarrow \delta_g \left(\Gamma^\gamma_{\alpha\beta} \right) = \frac{g^{\gamma\sigma}}{2} \left(\delta g_{\sigma\alpha;\beta} + \delta g_{\sigma\beta;\alpha} - \delta g_{\alpha\beta;\sigma} \right). \tag{319}$$

In the latter variation, we have made use of the fact that

$$\delta_{g_{\alpha\beta}} \left(g_{\kappa\lambda} \right) = \frac{1}{2} \left(\delta^\alpha_\kappa \delta^\beta_\lambda + \delta^\alpha_\lambda \delta^\beta_\kappa \right) \delta_g g_{\alpha\beta}. \tag{320}$$

Please note that the ";" (last term in last line in (319)) denotes the covariant derivative. Using the definition of the latter, the variation of the affine connection can further be evaluated:

$$\delta_g\left(\Gamma^\gamma_{\alpha\beta}\right) = \frac{1}{2}g^{\gamma\sigma}\left(\delta g_{\sigma\alpha;\beta} + \delta g_{\sigma\beta;\alpha} - \delta g_{\alpha\beta;\sigma}\right)$$

$$= \frac{g^{\gamma\sigma}}{4}\begin{pmatrix} 2\left(\delta g_{\sigma\alpha,\beta} + \delta g_{\sigma\beta,\alpha} - \delta g_{\alpha\beta,\sigma}\right) \\ -2\Gamma^\kappa_{\alpha\beta}g^\lambda_\sigma + \Gamma^\kappa_{\alpha\sigma}\left(g^\lambda_\beta - g^\lambda_\alpha\right) - 2\Gamma^\lambda_{\alpha\beta}g^\kappa_\sigma + \Gamma^\lambda_{\alpha\sigma}\left(g^\kappa_\beta - g^\kappa_\alpha\right) \end{pmatrix}\delta_g g_{\kappa\lambda}.$$

$$(321)$$

In the case of diagonal metrics and ignoring derivatives of the variation of the metric, we see that we can give a dramatically simplified formula for the variation of the affine connection $\delta_g\left(\Gamma^\gamma_{\alpha\beta}\right)$:

$$\delta_g\left(\Gamma^\gamma_{\alpha\beta}\right) = \frac{1}{2}\left[g^{\gamma\sigma}\delta_g\left(\ldots\right)_{\alpha\beta\sigma} + \left(\ldots\right)_{\alpha\beta\sigma}\delta_g g^{\gamma\sigma}\right]$$

$$= -\frac{1}{4}\left(\ldots\right)_{\alpha\beta\sigma}\left(g^{\gamma\kappa}g^{\sigma\lambda} + g^{\sigma\kappa}g^{\gamma\lambda}\right)\delta_g g_{\kappa\lambda}$$

$$= -\frac{1}{2}\left(g^{\gamma\kappa}\Gamma^\lambda_{\alpha\beta} + \Gamma^\kappa_{\alpha\beta}g^{\gamma\lambda}\right)\delta_g g_{\kappa\lambda}. \tag{322}$$

We will also need the variation of the Ricci tensor, which can be evaluated using (321) from

$$\delta_g\left(R_{\alpha\beta}\right) = \begin{pmatrix} -\delta_g\left[\Gamma^\sigma_{\beta\alpha}\right]_{,\alpha} + \delta_g\left[\Gamma^\sigma_{\alpha\beta}\right]_{,\sigma} \\ -\Gamma^\mu_{\sigma\alpha}\delta_g\Gamma^\sigma_{\beta\mu} - \Gamma^\sigma_{\beta\mu}\delta_g\Gamma^\mu_{\sigma\alpha} \\ +\Gamma^\mu_{\sigma\mu}\delta_g\Gamma^\sigma_{\alpha\beta} + \Gamma^\sigma_{\alpha\beta}\delta_g\Gamma^\mu_{\sigma\mu} \end{pmatrix} \equiv M^{\lambda\kappa}_{\alpha\beta}\delta_g g_{\kappa\lambda}, \tag{323}$$

which is giving us a rather complex expression with our current $\delta_g\left(\Gamma^\gamma_{\alpha\beta}\right)$. We therefore try to find a shorter form by starting from

$$\delta_g\Gamma^\gamma_{\alpha\beta} = \frac{1}{2}\delta_g\left[g^{\gamma\sigma}\left(g_{\sigma\alpha,\beta} + g_{\sigma\beta,\alpha} - g_{\alpha\beta,\sigma}\right)\right]$$

$$= \frac{\left(g_{\sigma\alpha,\beta} + g_{\sigma\beta,\alpha} - g_{\alpha\beta,\sigma}\right)}{2}\delta_g g^{\gamma\sigma} + \frac{g^{\gamma\sigma}\delta_g\left(g_{\sigma\alpha,\beta} + g_{\sigma\beta,\alpha} - g_{\alpha\beta,\sigma}\right)}{2}$$

$$= -\left(g^{\lambda\sigma}g^{\gamma\kappa} + g^{\lambda\gamma}g^{\sigma\kappa}\right)\frac{\left(g_{\sigma\alpha,\beta} + g_{\sigma\beta,\alpha} - g_{\alpha\beta,\sigma}\right)}{4}\delta_g g_{\kappa\lambda}$$

$$+ \frac{g^{\gamma\sigma}\left(\delta g_{\sigma\alpha,\beta} + \delta g_{\sigma\beta,\alpha} - \delta g_{\alpha\beta,\sigma}\right)}{2}$$

$$= -\frac{g^{\gamma\kappa}\Gamma^\lambda_{\alpha\beta} + g^{\gamma\lambda}\Gamma^\kappa_{\alpha\beta}}{2}\delta_g g_{\kappa\lambda} + \frac{g^{\gamma\sigma}\left(\delta g_{\sigma\alpha,\beta} + \delta g_{\sigma\beta,\alpha} - \delta g_{\alpha\beta,\sigma}\right)}{2}$$

$$= \frac{g^{\gamma\sigma}\left(\delta g_{\sigma\alpha,\beta} + \delta g_{\sigma\beta,\alpha} - \delta g_{\alpha\beta,\sigma}\right) - g^{\gamma\kappa}\Gamma^{\lambda}_{\alpha\beta}\delta g g_{\kappa\lambda} - g^{\gamma\lambda}\Gamma^{\kappa}_{\alpha\beta}\delta g g_{\kappa\lambda}}{2}$$

$$= \frac{g^{\gamma\sigma}\left(g^{\kappa}_{\alpha}g^{\lambda}_{\sigma}g^{\rho}_{\beta} + g^{\kappa}_{\sigma}g^{\lambda}_{\beta}g^{\rho}_{\alpha} - g^{\kappa}_{\alpha}g^{\lambda}_{\beta}g^{\rho}_{\sigma}\right)\delta g_{\kappa\lambda,\rho} - \left(g^{\gamma\kappa}\Gamma^{\lambda}_{\alpha\beta} + g^{\gamma\lambda}\Gamma^{\kappa}_{\alpha\beta}\right)\delta g_{\kappa\lambda}}{2}.$$

$$(324)$$

With respect to the ordinary derivative of the variated metric ($\delta g_{\sigma\alpha,\beta} + \delta g_{\sigma\beta,\alpha} - \delta g_{\alpha\beta,\sigma}$), we use the following abbreviation:

$$\left(g^{\kappa}_{\alpha}g^{\lambda}_{\sigma}g^{\rho}_{\beta} + g^{\kappa}_{\sigma}g^{\lambda}_{\beta}g^{\rho}_{\alpha} - g^{\kappa}_{\alpha}g^{\lambda}_{\beta}g^{\rho}_{\sigma}\right) \equiv GX^{\kappa\lambda\rho}_{\alpha\beta\sigma}. \qquad (325)$$

This allows us to give the following compact form for the variation of the Ricci tensor:

$$\delta_g R_{\alpha\beta} = \frac{1}{2}\left(\begin{array}{l}
-\left[\begin{array}{l}\left(\left(g^{\sigma\rho}GX^{\kappa\lambda\eta}_{\sigma\beta\rho}\right)_{,\alpha} - g^{\eta}_{\alpha}\left(\Gamma^{\kappa}_{\sigma\beta}g^{\sigma\lambda} + \Gamma^{\lambda}_{\sigma\beta}g^{\sigma\kappa}\right)\right)\delta g_{\kappa\lambda,\eta} \\ +\left(g^{\omega}_{\alpha}g^{\sigma\rho}GX^{\kappa\lambda\eta}_{\sigma\beta\rho}\right)\delta g_{\kappa\lambda,\eta\omega} - \left(\Gamma^{\kappa}_{\sigma\beta}g^{\sigma\lambda} + \Gamma^{\lambda}_{\sigma\beta}g^{\sigma\kappa}\right)_{,\alpha}\delta g_{\kappa\lambda}\end{array}\right] \\
+\left[\begin{array}{l}\left(\left(g^{\sigma\rho}GX^{\kappa\lambda\eta}_{\alpha\beta\rho}\right)_{,\sigma} - g^{\eta}_{\sigma}\left(\Gamma^{\kappa}_{\sigma\beta}g^{\sigma\lambda} + \Gamma^{\lambda}_{\sigma\beta}g^{\sigma\kappa}\right)\right)\delta g_{\kappa\lambda,\eta} \\ +\left(g^{\omega\rho}GX^{\kappa\lambda\eta}_{\alpha\beta\rho}\right)\delta g_{\kappa\lambda,\eta\omega} - \left(\Gamma^{\kappa}_{\alpha\beta}g^{\sigma\lambda} + \Gamma^{\lambda}_{\alpha\beta}g^{\sigma\kappa}\right)_{,\sigma}\delta g_{\kappa\lambda}\end{array}\right] \\
-\Gamma^{\sigma}_{\mu\alpha}\left(g^{\mu\rho}GX^{\kappa\lambda\eta}_{\sigma\beta\rho}\delta g_{\kappa\lambda,\eta} - \left(\Gamma^{\kappa}_{\sigma\beta}g^{\mu\lambda} + \Gamma^{\lambda}_{\sigma\beta}g^{\mu\kappa}\right)\delta g_{\kappa\lambda}\right) \\
-\Gamma^{\mu}_{\beta\sigma}\left(g^{\sigma\rho}GX^{\kappa\lambda\eta}_{\alpha\mu\rho}\delta g_{\kappa\lambda,\eta} - \left(\Gamma^{\kappa}_{\alpha\mu}g^{\sigma\lambda} + \Gamma^{\lambda}_{\alpha\mu}g^{\sigma\kappa}\right)\delta g_{\kappa\lambda}\right) \\
+\Gamma^{\mu}_{\sigma\mu}\left(g^{\sigma\rho}GX^{\kappa\lambda\eta}_{\alpha\beta\rho}\delta g_{\kappa\lambda,\eta} - \left(\Gamma^{\kappa}_{\alpha\beta}g^{\sigma\lambda} + \Gamma^{\lambda}_{\alpha\beta}g^{\sigma\kappa}\right)\delta g_{\kappa\lambda}\right) \\
+\Gamma^{\sigma}_{\alpha\beta}\left(g^{\mu\rho}GX^{\kappa\lambda\eta}_{\sigma\mu\rho}\delta g_{\kappa\lambda,\eta} - \left(\Gamma^{\kappa}_{\mu\sigma}g^{\mu\lambda} + \Gamma^{\lambda}_{\mu\sigma}g^{\mu\kappa}\right)\delta g_{\kappa\lambda}\right)
\end{array}\right)$$

$$\equiv Mn^{\lambda\kappa}_{\alpha\beta}\delta_g g_{\kappa\lambda} + \tilde{M}n^{\kappa\lambda\eta}_{\alpha\beta}\delta g_{\kappa\lambda,\eta} + \hat{M}n^{\kappa\lambda\eta\omega}_{\alpha\beta}\delta g_{\kappa\lambda,\eta\omega}. \qquad (326)$$

A further simplification is possible for the assumption that all derivatives of the metric variation $\delta g_{\alpha\beta}$ are small compared to the variation itself, which gives*

$$\delta_g\left(R_{\alpha\beta}\right) \Rightarrow -\frac{1}{2}\left[\begin{array}{l}\left(\begin{array}{l}-\left[g^{\sigma\kappa}\Gamma^{\lambda}_{\beta\sigma}\right]_{,\alpha} + \left[g^{\sigma\kappa}\Gamma^{\lambda}_{\alpha\beta}\right]_{,\sigma} \\ -\Gamma^{\sigma}_{\mu\alpha}g^{\mu\kappa}\Gamma^{\lambda}_{\sigma\beta} - \Gamma^{\mu}_{\beta\sigma}g^{\sigma\kappa}\Gamma^{\lambda}_{\alpha\mu} \\ +\Gamma^{\mu}_{\sigma\mu}g^{\sigma\kappa}\Gamma^{\lambda}_{\alpha\beta} + \Gamma^{\sigma}_{\alpha\beta}g^{\mu\kappa}\Gamma^{\lambda}_{\sigma\mu}\end{array}\right) \\
+\left(\begin{array}{l}-\left[g^{\sigma\lambda}\Gamma^{\kappa}_{\beta\sigma}\right]_{,\alpha} + \left[g^{\sigma\lambda}\Gamma^{\kappa}_{\alpha\beta}\right]_{,\sigma} \\ -\Gamma^{\sigma}_{\mu\alpha}g^{\mu\lambda}\Gamma^{\kappa}_{\sigma\beta} - \Gamma^{\mu}_{\beta\sigma}g^{\sigma\lambda}\Gamma^{\kappa}_{\alpha\mu} \\ +\Gamma^{\mu}_{\sigma\mu}g^{\sigma\lambda}\Gamma^{\kappa}_{\alpha\beta} + \Gamma^{\sigma}_{\alpha\beta}g^{\mu\lambda}\Gamma^{\kappa}_{\sigma\mu}\end{array}\right)\end{array}\right]\delta_g g_{\kappa\lambda} \qquad (327)$$

---

*This simplified expression is not covariant anymore. Thus, it should be used with care.

where we have used the identities already given in (319) and (321). Due to the assumed shear-freedom of the metric, we are going to consider here, the symmetry is also guaranteed by the two shorter and simpler forms

$$\delta_g \left(R_{\alpha\beta} \right) \Rightarrow - \begin{pmatrix} - \left[g^{\sigma\kappa} \Gamma^\lambda_{\beta\sigma} \right]_{,\alpha} + \left[g^{\sigma\kappa} \Gamma^\lambda_{\alpha\beta} \right]_{,\sigma} \\ -\Gamma^\sigma_{\mu\alpha} g^{\mu\kappa} \Gamma^\lambda_{\sigma\beta} - \Gamma^\mu_{\beta\sigma} g^{\sigma\kappa} \Gamma^\lambda_{\alpha\mu} \\ +\Gamma^\mu_{\sigma\mu} g^{\sigma\kappa} \Gamma^\lambda_{\alpha\beta} + \Gamma^\sigma_{\alpha\beta} g^{\mu\kappa} \Gamma^\lambda_{\sigma\mu} \end{pmatrix} \delta_g g_{\kappa\lambda}$$

$$\delta_g \left(R_{\alpha\beta} \right) \Rightarrow - \begin{pmatrix} - \left[g^{\sigma\lambda} \Gamma^\kappa_{\beta\sigma} \right]_{,\alpha} + \left[g^{\sigma\lambda} \Gamma^\kappa_{\alpha\beta} \right]_{,\sigma} \\ -\Gamma^\sigma_{\mu\alpha} g^{\mu\lambda} \Gamma^\kappa_{\sigma\beta} - \Gamma^\mu_{\beta\sigma} g^{\sigma\lambda} \Gamma^\kappa_{\alpha\mu} \\ +\Gamma^\mu_{\sigma\mu} g^{\sigma\lambda} \Gamma^\kappa_{\alpha\beta} + \Gamma^\sigma_{\alpha\beta} g^{\mu\lambda} \Gamma^\kappa_{\sigma\mu} \end{pmatrix} \delta_g g_{\kappa\lambda}$$

(328)

where we have used the identities already given in (319) and (322).

It shall be noted that most of the simple metrics we consider here do not need the approximation (327). There the GX-terms will fall out anyway, but we are still using it within later equations, because of brevity. The reader should bear in mind, however, that for the general case always equation (326) needs to be used (see [40] for evaluations and additional discussion below).

Now we are able to evaluate the integrand in (317) and by setting the classical matter term identical to zero we have

$$\delta_g W = 0 = \delta_g \int_V d^n x \sqrt{-g} R = \delta_g \int_V d^n x \sqrt{-g} g^{\alpha\beta} R_{\alpha\beta}$$

$$= \int_V d^n x \delta_g \left(\sqrt{-g} g^{\alpha\beta} R_{\alpha\beta} \right)$$

$$= \int_V d^n x \left[\delta_g \left(\sqrt{-g} g^{\alpha\beta} \right) R_{\alpha\beta} + \sqrt{-g} g^{\alpha\beta} \delta_g \left(R_{\alpha\beta} \right) \right]$$

$$= \int_V d^n x \sqrt{-g} \left[-G^{\kappa\lambda} \delta_g g_{\kappa\lambda} + g^{\alpha\beta} \delta_g R_{\alpha\beta} \right],$$

(329)

with $G^{\alpha\beta} = R^{\alpha\beta} - \frac{1}{2} R g^{\alpha\beta}$ denoting the Einstein tensor. It has to be noted that the result does not change when we add a constant Λ, giving us $G^{\alpha\beta} + \Lambda g^{\alpha\beta} = R^{\alpha\beta} - \frac{1}{2} R g^{\alpha\beta} + \Lambda g^{\alpha\beta}$ (c.f. (318)).

It was assumed by Hilbert [4] that the second term in the last line in the integrand of (329), which is to say the term $g^{\alpha\beta} \delta_g R_{\alpha\beta}$, could be made a surface integral and thus, disregarded. Thereby local geodesic coordinates and the option of interchanging variation and differentiation were applied. These measures, however, require the basic assumption of a continuous space-time. Now assuming that the space is substructured, the present author comes to the conclusion that the disregard of the term $g^{\alpha\beta}\delta_g R_{\alpha\beta}$ is not only not always valid, but that it actually contains the quantum fields and matter. Assuming the space in question to be substructured or granular, we have to substitute the technique of the surface integral by sum. This sum requires a certain number of summands to actually sum up to zero. Thus, we might conclude that only from a certain scale upwards this sum would truly lead to a zero-sum for the $g^{\alpha\beta}\delta_g R_{\alpha\beta}$ term. Underneath that scale, we should take care about the conditions this term adds to the classical Einstein field equations.

One might put it like this:

Mathematically, Hilbert was perfectly right to ban the $g^{\alpha\beta}\delta_g R_{\alpha\beta}$ part of his variation integral from the solution, because, in fact, that very integral gives zero if—and that is the important point here—if integrated. So, Hilbert has found, as Einstein had done with slightly other, which is to say less rigorous means, ONE possible outcome of his variation problem. Before this integration inside the Hilbert variation is performed, however, there are certain things possible for the integrand. For instance, there is no law fixing the surface over which the surface integration, resulting from the $g^{\alpha\beta}\delta_g R_{\alpha\beta}$ term, has to be performed. There is no fixing about how the integrand could look like, before actually being summed up in the integral and forming a zero sum. Thus, when and where the integral—after integration—has to become a zero sum is a degree of freedom. Obviously, nobody had told the universe not to use this degree of freedom and nobody has told if off for actually using it in the end neither. Obviously, the universe made use of this opportunity quite merrily and added it to its own toolbox for evolution.

For illustration: The reader might perhaps imagine a huge black board on which a mathematical task like "1 minus 1" is been written, also in huge letters. From a far distance an observers would see both numbers plus the minus sign in between and could perform the evaluation, resulting at the usual $1 - 1 = 0$. Coming closer and closer, however, the observer would not be able to see the complete task and therefore could not bring the two numbers together in order to form the familiar zero outcome. In other words, it seems that it depends on the scale whether a certain evaluation can be done or not. Of course, this is just another way of introducing scale relativity or fractal dimensions. Only that, this time, we leave the spaces intact, meaning continuous and instead of considering them to become grainy at a certain—smaller—scale, we rather subject the possibility of executing certain continuous evaluations, performed on these very spaces, to a scale-dependency. This leaves the classical solutions at bigger scales intact, but adds "add-ons" the moment we consider smaller and smaller portions of the universe, which is to say moving closer and closer towards the universal black board.

In conclusion: The universe does not need to be grainy in order to obtain the quantum theory in harmonic unison with relativity. No, we only need to add the "condition or boundary of scale" to each and every continuous evaluation, being performed on space-time.

Having uttered that suspicion, we only need to show now that these intrinsic degrees of freedom residing in the Einstein–Hilbert action and originally being discarded by both scientists, are giving us what is known as quantum theory.

However, as we already have resorted to rather esoteric things like "allow the universe" to choose to "integrate" (or surface integrate) the Ricci tensor term or to leave it within the integrand treating it as differential expression, we should also—in principle—allow for certain choices with respect to the ratio of integration to differentiation of the $g^{\alpha\beta}\delta_g R_{\alpha\beta}$ term. After all, there seems to be a great variety of options to treat this term. Here we will therefore only concentrate on "obviously reasonable" ratios, leading to equations similar to the ones we have already obtained in this book in connection with the technique of the "intelligent zero" (c.f. the previous chapters). This way we will demonstrate that apparently the quantum theory is hidden inside the Einstein–Hilbert action. We do not intend to give a complete discussion on all options coming with the various ways the $g^{\alpha\beta}\delta_g R_{\alpha\beta}$ term could be treated. This task is either for a follow-up book to this one or will be performed in our small series of "Einstein had it" (e.g. [38–40]).

One important aspect, however, should not be left out here. Namely, when investigating the additional term $g^{\alpha\beta}\delta_g R_{\alpha\beta}$, we find that our familiar quantum effects only occur if an additional negative sign is added to $g^{\alpha\beta}\delta_g R_{\alpha\beta}$ or its evaluated form (326). One even finds that there is a variety of factors the term could be multiplied with, which do give quite reasonable solutions. While the direct evaluation of $g^{\alpha\beta}\delta_g R_{\alpha\beta}$ does only allow for one fixed version of the variated Ricci tensor, we should ask ourselves why also forms like $A \times g^{\alpha\beta}\delta_g R_{\alpha\beta}$ can give certain reasonable solutions. The only straight forward and reasonable explanation for this was found in the fact that our universe might be embedded in a higher dimensional space and that in fact the surface term from this embedding space (meaning the term $\gamma^{ab}\delta_\gamma R_{ab}$, with γ^{ab} denoting the higher dimensional metric) could be acting within our universe and thus, effectively, changes sign and coefficient of "our" $g^{\alpha\beta}\delta_g R_{\alpha\beta}$. In fact, it will be shown elsewhere that such a "coupling in" from upper dimensions is even possible as subsurface term from "two levels higher up." One might even, rather boldly, conclude that connection or "entanglement" of the various spaces or parts of spaces is what we encounter as our matter. As this elaboration needs quite some space and—for completeness—also requires the consideration of higher orders of variation, it will be performed in [40].

Just in order to give a hint, however, how the evaluation has to be performed, we transform (326) into a form with only covariant derivatives for the metric variations, which gives us

$$
\delta_g R_{\alpha\beta} = \frac{1}{2}
\begin{pmatrix}
g^{\sigma\rho}GX^{\kappa\lambda\eta}_{\alpha\beta\rho}g^{\omega}_{\alpha}\delta g_{\kappa\lambda;\eta\omega} + g^{\omega\rho}GX^{\kappa\lambda\eta}_{\alpha\beta\rho}\delta g_{\kappa\lambda;\eta\omega} \\[4pt]
- \left[\left[g^{\sigma\rho}GX^{\kappa\lambda\eta}_{\sigma\beta\rho} \right]_{,\alpha} + g^{\sigma\rho}GX^{\kappa\lambda\eta}_{\sigma\beta\rho}\left(\Gamma^{\varpi}_{\kappa\alpha}g^{\kappa}_{\varpi} + \Gamma^{\varpi}_{\lambda\alpha}g^{\lambda}_{\varpi} + \Gamma^{\varpi}_{\eta\alpha}g^{\eta}_{\varpi} \right) \right] \delta g_{\kappa\lambda;\eta} \\[4pt]
+ \left[\left[g^{\sigma\rho}GX^{\kappa\lambda\eta}_{\alpha\beta\rho} \right]_{,\sigma} + g^{\sigma\rho}GX^{\kappa\lambda\eta}_{\alpha\beta\rho}\left(\Gamma^{\varpi}_{\kappa\sigma}g^{\kappa}_{\varpi} + \Gamma^{\varpi}_{\lambda\sigma}g^{\lambda}_{\varpi} + \Gamma^{\varpi}_{\eta\sigma}g^{\eta}_{\varpi} \right) \right] \delta g_{\kappa\lambda;\eta} \\[4pt]
+ \left[\begin{array}{l} -\Gamma^{\sigma}_{\mu\alpha}g^{\mu\rho}GX^{\kappa\lambda\eta}_{\sigma\beta\rho} - \Gamma^{\mu}_{\beta\sigma}g^{\sigma\rho}GX^{\kappa\lambda\eta}_{\alpha\mu\rho} \\ +\Gamma^{\mu}_{\sigma\mu}g^{\sigma\rho}GX^{\kappa\lambda\eta}_{\alpha\beta\rho} + \Gamma^{\sigma}_{\alpha\beta}g^{\mu\rho}GX^{\kappa\lambda\eta}_{\sigma\mu\rho} \end{array} \right] \delta g_{\kappa\lambda;\eta}
\end{pmatrix}.
$$
(330)

We see how the variation of the Ricci tensor is consisting of divergences of the variation of the metric. Interestingly, the $\delta_g g_{\kappa\lambda}$ terms have disappeared. Now we can show that integration by parts does not only allow us to get rid of the metric variation derivative terms $\delta g_{\kappa\lambda;\eta}$, $\delta g_{\kappa\lambda;\eta\omega}$, but also directly provides us with the required "−" sign. In addition, we find that integration by parts twice allows for further surface terms from spaces of higher dimensions coupling into our space-time. In result, we end up with quite some "flexibility" regarding the set of equations known as Einstein field equations and this—we need to emphasize it here—is without any change of the linear R-Lagrangian being put into the Einstein–Hilbert action.

For more the reader is referred to [40]. As a consequence, however, here we will concentrate on situations with $A = -1$. Thus, for demonstration, the assumed to be missing "−" will be added to all our $g^{\alpha\beta}\delta_g R_{\alpha\beta}$ matter terms throughout the chapter without further notice.

For a start, we will demonstrate this by using the intrinsic degrees of freedom residing in the terms neglected by Einstein and Hilbert and demand the following:

$$\delta_g\left(R_{\alpha\beta}\right) = 0 \Rightarrow \begin{cases} -\left[g^{\sigma\kappa}\Gamma^\lambda_{\beta\sigma}\right]_{,\alpha} - \Gamma^\sigma_{\mu\alpha}g^{\mu\kappa}\Gamma^\lambda_{\sigma\beta} + \Gamma^\mu_{\sigma\mu}g^{\sigma\kappa}\Gamma^\lambda_{\alpha\beta} = 0 \\ \left[g^{\sigma\kappa}\Gamma^\lambda_{\alpha\beta}\right]_{,\sigma} - \Gamma^\mu_{\beta\sigma}g^{\sigma\kappa}\Gamma^\lambda_{\alpha\mu} + \Gamma^\sigma_{\alpha\beta}g^{\mu\kappa}\Gamma^\lambda_{\sigma\mu} = 0 \end{cases}, \quad (331)$$

or

$$\delta_g\left(R_{\alpha\beta}\right) = 0 \Rightarrow \begin{cases} -\left[g^{\sigma\kappa}\Gamma^\lambda_{\beta\sigma}\right]_{,\alpha} - \Gamma^\sigma_{\mu\alpha}g^{\mu\kappa}\Gamma^\lambda_{\sigma\beta} - \Gamma^\mu_{\beta\sigma}g^{\sigma\kappa}\Gamma^\lambda_{\alpha\mu} = 0 \\ \left[g^{\sigma\kappa}\Gamma^\lambda_{\alpha\beta}\right]_{,\sigma} + \Gamma^\mu_{\sigma\mu}g^{\sigma\kappa}\Gamma^\lambda_{\alpha\beta} + \Gamma^\sigma_{\alpha\beta}g^{\mu\kappa}\Gamma^\lambda_{\sigma\mu} = 0 \end{cases}. \quad (332)$$

The One-Dimensional Case

It is convenient to observe this by going to the one-dimensional case, where for brevity we set

$$g_{\sigma\kappa} \to g; \quad g^{\sigma\kappa} \to \frac{1}{g}; \quad g_{\sigma\kappa,\alpha} \to g'. \quad (333)$$

Then we obtain from (327)

$$\delta_g\left(R_{\alpha\beta}\right) = \left(-\left[\frac{g'}{2g^2}\right]' + \left[\frac{g'}{2g^2}\right]' - 2\left(\frac{g'}{2g^2}\right)^2 + 2\left(\frac{g'}{2g^2}\right)^2\right)\delta_g g. \quad (334)$$

No matter the fact that this already gives zero, but simply by using the intrinsic degree of freedom, which this equation contains, we could still demand a solution of the form

$$\left[\frac{g'}{2g^2}\right]' - 2\left(\frac{g'}{2g^2}\right)^2 = 0. \quad (335)$$

Evaluation of the above gives

$$-\frac{g'' \cdot g^2 - 2 \cdot g \cdot (g')^2}{2 \cdot g^4} - \frac{2}{4}\cdot\left(\frac{g'}{g^2}\right)^2 = -\frac{g'' \cdot g - 2 \cdot (g')^2}{2 \cdot g^3} - \left(\frac{g'}{g}\right)^2\cdot\frac{1}{2 \cdot g}$$

$$= -g'' \cdot g + 2 \cdot (g')^2 - (g')^2 = -g'' \cdot g + (g')^2 = 0 \quad (336)$$

which gives the general solution

$$g[x] = C \cdot e^{\mu \cdot x}; \quad \mu \in \mathbb{C} \quad (337)$$

and structurally resembles both the Schrödinger and the Dirac equation for special potential settings, because we can write (336) as two equations like

$$0 = g'' - \mu^2 \cdot g$$

$$0 = -\mu^2 \cdot g^2 + (g')^2 \Rightarrow \begin{cases} i \cdot \mu \cdot g + g' \\ i \cdot \mu \cdot g - g' \end{cases} = 0; \quad i = \sqrt{-1}, \quad (338)$$

where we have applied the vectorial root decomposition from the previous sections again (alternatively see also [9]).

Comparison of the first equation in (338) with the classical Schrödinger equation

$$\left[-i \cdot \hbar \cdot \partial_t - \frac{\hbar^2}{2 \cdot m} \Delta_{\text{Schrödinger}} + V \right] \Psi = 0, \tag{339}$$

thereby reducing the latter to one spatial dimension and making it time-independent (after all, we are considering a one-dimensional Einstein–Hilbert action),

$$-\frac{\hbar^2}{2 \cdot m} \Psi'' + V \cdot \Psi = E_n \cdot \Psi, \tag{340}$$

with the energy eigenvalues E_n gives us the parameter μ:

$$\mu^2 = \frac{2 \cdot m}{\hbar^2} \tag{341}$$

in the case of a potential $V - E_n = 1$. In the case of $V = V[x]$ things are getting more complicated, because, as before with the technique of the intelligent zero, we see that the potential $V[x]$ is taken on by the metric. This time, however, meaning by deriving the quantum equations directly from the Einstein–Hilbert action, we have the metric alone and do not even need an auxiliary function as in our previous sections, where we had used the technique of the "intelligent zeros." By the means of (338) and (340) any arbitrary—classical potential $V[x]$ should now be expressed by its "metric distortion equivalent" (c.f. [16]). So, using the general solution (337) for a given classical potential V in the Schrödinger picture, the corresponding metric could be found by

$$-\frac{\hbar^2}{2 \cdot m} \Psi'' + (V[x] - E_n) \cdot \Psi = 0 = -\frac{g''}{\mu^2} + g \Rightarrow V \cdot \Psi = g$$

$$\Rightarrow g[x] = \int_{-\infty}^{\infty} F[\mu] \cdot e^{2 \cdot \pi \cdot i \cdot \mu \cdot x} d\mu + \int_{0}^{\infty} L[\mu] \cdot e^{-\mu \cdot x} d\mu$$

$$F[\mu] = \int_{-\infty}^{\infty} (V[x] - E_n) \cdot \Psi[x] \cdot e^{-2 \cdot \pi \cdot i \cdot \mu \cdot x} dx$$

$$L[\mu] = \frac{1}{2\pi i} \int_{\gamma - i \cdot T}^{\gamma - i \cdot T} (V[x] - E_n) \cdot \Psi[x] \cdot e^{\mu \cdot x} dx. \tag{342}$$

For convenience, we also write the above for the case of the Klein–Gordon equation in 1D:

$$\left[\partial_x^2 - \frac{m^2 c^2}{\hbar^2} + V[x] - E_n\right] \Psi_{KG} = 0 = g'' - \mu^2 g$$

$$\Rightarrow \left[\frac{m^2 c^2}{\hbar^2} - V[x] + E_n\right] \Psi_{KG} = \mu^2 g \Rightarrow$$

$$g[x] = \int_{-\infty}^{\infty} F[\mu] \cdot e^{2\cdot\pi\cdot i\cdot\mu\cdot x} d\mu + \int_0^{\infty} L[\mu] \cdot e^{-\mu\cdot x} d\mu$$

$$F[\mu] = \int_{-\infty}^{\infty} \left(\frac{m^2 c^2}{\hbar^2} - V[x] + E_n\right) \cdot \Psi_{KG}[x] \cdot e^{-2\cdot\pi\cdot i\cdot\mu\cdot x} dx$$

$$L[\mu] = \frac{1}{2\pi i} \int_{\gamma-i\cdot T}^{\gamma-i\cdot T} \left(\frac{m^2 c^2}{\hbar^2} - V[x] + E_n\right) \cdot \Psi_{KG}[x] \cdot e^{\mu\cdot x} dx \quad (343)$$

and the Dirac equation:

$$\left(\gamma_{1D}\partial_x + i \cdot \frac{m \cdot c}{\hbar} + V - E_n\right) \Psi_D = 0 = \left\{\begin{matrix} i \cdot \mu \cdot g + g' \\ i \cdot \mu \cdot g - g' \end{matrix}\right\}; \quad \gamma_{1D} = \pm 1$$

$$\Rightarrow \left(i \cdot \frac{m \cdot c}{\hbar} + V - E_n\right) \Psi_D = i \cdot \mu \cdot g$$

$$\Rightarrow g[x] = \int_{-\infty}^{\infty} F[\mu] \cdot e^{2\cdot\pi\cdot i\cdot\mu\cdot x} d\mu + \int_0^{\infty} L[\mu] \cdot e^{-\mu\cdot x} d\mu$$

$$F[\mu] = \int_{-\infty}^{\infty} \left(i \cdot \frac{m \cdot c}{\hbar} + V - E_n\right) \cdot \Psi_D[x] \cdot e^{-2\cdot\pi\cdot i\cdot\mu\cdot x} dx$$

$$L[\mu] = \frac{1}{2\pi i} \int_{\gamma-i\cdot T}^{\gamma-i\cdot T} \left(i \cdot \frac{m \cdot c}{\hbar} + V - E_n\right) \cdot \Psi_D[x] \cdot e^{\mu\cdot x} dx. \quad (344)$$

The Harmonic Quantum Oscillator in 1D in the Metric Picture

Now we want to investigate an example, which shall illustrate us how a classical potential and its subsequent quantum states influence the metric of a given space. Therefore, we are going to make use of our previous results of the "The One-Dimensional Case." As it is so popular, we here

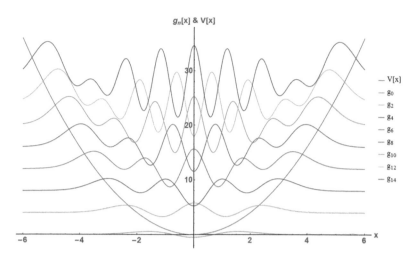

g_n[x] & V[x]

Figure 42 Metric distortion of a harmonic oscillator for the first even eigenstates. For better orientation, we have also drawn the classical potential $V[x]$.

consider the harmonic oscillator, and applying Eq. (342) we are going to evaluate the resulting metric for a variety of energy states of the harmonic oscillator. With the potential $V[x] \sim x^2$, we have the classical solutions of the harmonic oscillator known to be (see section "The Classical Harmonic Quantum Oscillator within the Metric Picture, or the Theory of Everything"):

$$\psi_n(x) := \left(\frac{m \cdot \omega}{\pi \cdot \hbar}\right)^{\frac{1}{4}} \frac{1}{\sqrt{2^n n!}} H_n[w] * e^{\left[\frac{-w^2}{2}\right]}; \quad w = x \cdot \sqrt{\frac{m \cdot \omega}{\hbar}}. \quad (345)$$

Applying (342) and setting $m = \hbar = \omega = 1$ we find the metric solutions for the first eigenstates in Figs. 42 and 43.

Applying the integral

$$\int_{x_0}^{x_0+x} \sqrt{g_{xx}} dx' = \int_{x_0}^{x_0+x} \sqrt{F[x']} dx', \quad (346)$$

we can also evaluate the proper distance along our coordinate x. There we do not only find that the space "propagates" in steps (Figs. 44 and 45) but also brings about a new coordinate (or property or dimension) of imaginary character (Figs. 46 and 47). We might interpret this new dimension as time (Fig. 48), because we see that within the center region of the harmonic oscillator this part of the integrated line element behaves linear and its square becomes negative (just like time does in the General Theory of

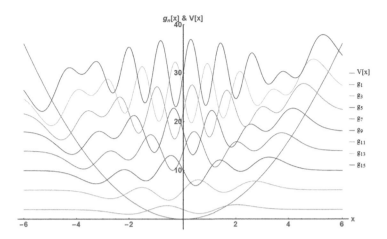

Figure 43 Metric distortion of a harmonic oscillator for the first odd eigenstates. For better orientation, we have also drawn the classical potential $V[x]$.

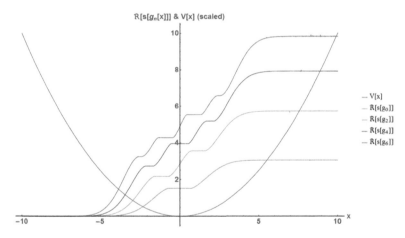

Figure 44 Real part of line element integral (distance) as found from (346) of a harmonic oscillator for the first even eigenstates. For better orientation, we have also drawn the classical potential $V[x]$.

Relativity). Taking more eigenstates into consideration, increases the linear area for the new imaginary dimension our originally one-dimensional space has born the moment it has taken on energy. One also discovers that the proper space, an intrinsic particle or observer could travel within, of our one

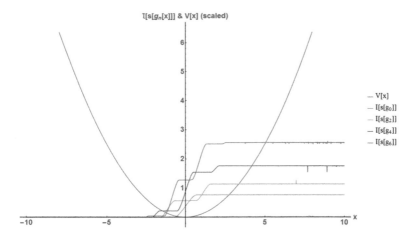

Figure 45 Imaginary part of line element integral (distance) as found from (346) of a harmonic oscillator for the first even eigenstates. For better orientation, we have also drawn the classical potential $V[x]$.

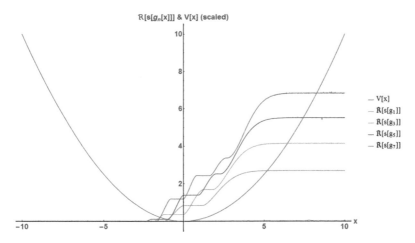

Figure 46 Real part of line element integral (distance) as found from (346) of a harmonic oscillator for the first odd eigenstates. For better orientation, we have also drawn the classical potential $V[x]$.

dimensional space increases with the order of the eigenstates and thus, the energy. We therefore conclude: the higher the energy, the bigger the space holding this energy or

$$\text{Space}(-\text{Time}) = \text{Energy}$$

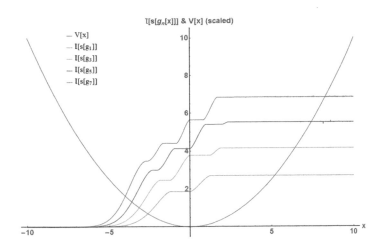

Figure 47 Imaginary part of line element integral (distance) as found from (346) of a harmonic oscillator for the first odd eigenstates. For better orientation, we have also drawn the classical potential $V[x]$.

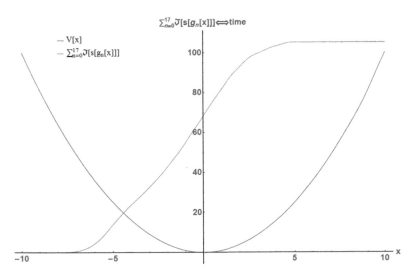

Figure 48 Imaginary part of line element integral (distance) as found from (346) of a harmonic oscillator for the sum of the first 17 eigenstates. We see that within the center region of the harmonic oscillator this part of the integrated line element behaves like time. Taking more eigenstates into consideration, increases the linear area for the new imaginary dimension, our originally one-dimensional space has given birth to the moment it has taken on energy. For better orientation, we have also drawn the classical potential $V[x]$.

The Three-Dimensional Case

We make a metric approach of the form

$$g_{\sigma\kappa} = \begin{pmatrix} X\,[x] & 0 & 0 \\ 0 & Y\,[y] & 0 \\ 0 & 0 & Z\,[z] \end{pmatrix}. \tag{347}$$

We evaluate all summands in the variation of the Ricci tensor in order to find out about possible intrinsic degrees of freedom (c.f. (327)):

$$-\left[g^{\sigma\kappa}\Gamma^\lambda_{\beta\sigma}\right]_{,\alpha} = \text{Trace}\left(\frac{-2X'^2 + X\cdot X''}{2X^4},\ \frac{-2Y'^2 + Y\cdot Y''}{2Y^4},\ \frac{-2Z'^2 + Z\cdot Z''}{2Z^4}\right) \tag{348}$$

$$\left[g^{\sigma\kappa}\Gamma^\lambda_{\alpha\beta}\right]_{,\sigma} = \begin{pmatrix} \frac{2X'^2 - X\cdot X''}{2X^4} & \frac{X'\cdot Y'}{2X^2\cdot Y^2} & \frac{X'\cdot Z'}{2X^2\cdot Z^2} \\ \frac{X'\cdot Y'}{2X^2\cdot Y^2} & \frac{2Y'^2 - Y\cdot Y''}{2Y^4} & \frac{Y'\cdot Z'}{2Y^2\cdot Z^2} \\ \frac{X'\cdot Z'}{2X^2\cdot Z^2} & \frac{Y'\cdot Z'}{2Y^2\cdot Z^2} & \frac{2Z'^2 - Z\cdot Z''}{2Z^4} \end{pmatrix} \tag{349}$$

$$-\Gamma^\mu_{\sigma\alpha}g^{\sigma\kappa}\Gamma^\lambda_{\beta\mu} = -\Gamma^\sigma_{\beta\mu}g^{\mu\kappa}\Gamma^\lambda_{\sigma\alpha} = -\Gamma^\sigma_{\alpha\beta}g^{\rho\kappa}\Gamma^\lambda_{\sigma\rho} = \text{Trace}\left(\frac{X'^2}{4X^4},\ \frac{Y'^2}{4Y^4},\ \frac{Z'^2}{4Z^4}\right) \tag{350}$$

$$\Gamma^\rho_{\sigma\rho}g^{\sigma\kappa}\Gamma^\lambda_{\alpha\beta} = -\frac{1}{4}\begin{pmatrix} \frac{X'^2}{X^4} & \frac{X'\cdot Y'}{X^2\cdot Y^2} & \frac{X'\cdot Z'}{X^2\cdot Z^2} \\ \frac{X'\cdot Y'}{X^2\cdot Y^2} & \frac{Y'^2}{Y^4} & \frac{Y'\cdot Z'}{Y^2\cdot Z^2} \\ \frac{X'\cdot Z'}{X^2\cdot Z^2} & \frac{Y'\cdot Z'}{Y^2\cdot Z^2} & \frac{Z'^2}{Z^4} \end{pmatrix} \tag{351}$$

where, for brevity, we have defined

$$T' = \partial_t T\,[t] = \frac{\partial T\,[t]}{\partial t};\ X' = \partial_x X\,[x];\ Y' = \partial_y Y\,[y];\ Z' = \partial_z Z\,[z]. \tag{352}$$

The simplest intrinsic solution would be constants like $X\,[x] = C_x$, $Y\,[y] = C_y$ and $Z\,[z] = C_z$. Thus, with $C_x = C_y = C_z = 1$ we would have a flat space solution without any quantum effects. However, we could also obtain this from sums of solutions, resulting from certain combinations of the terms above and demanding them to become zero. Thus, we construct the following intrinsic equations:

$$\delta_g\left(R_{\alpha\beta}\right) = 0 \Rightarrow \begin{cases} -\left[g^{\sigma\kappa}\Gamma^\lambda_{\beta\sigma}\right]_{,\alpha} - \Gamma^\mu_{\sigma\alpha}g^{\sigma\kappa}\Gamma^\lambda_{\beta\mu} - \Gamma^\sigma_{\beta\mu}g^{\mu\kappa}\Gamma^\lambda_{\sigma\alpha} = 0 \\ \left[g^{\sigma\kappa}\Gamma^\lambda_{\alpha\beta}\right]_{,\sigma} + \Gamma^\sigma_{\alpha\beta}g^{\rho\kappa}\Gamma^\lambda_{\sigma\rho} + \Gamma^\rho_{\sigma\rho}g^{\sigma\kappa}\Gamma^\lambda_{\alpha\beta} = 0 \end{cases}$$

$$= \begin{cases} -\text{Trace}\left(\frac{X'^2 - X\cdot X''}{2X^4},\ \frac{Y'^2 - Y\cdot Y''}{2Y^4},\ \frac{Z'^2 - Z\cdot Z''}{2Z^4}\right) = 0 \\ \frac{1}{2}\cdot\begin{pmatrix} \frac{X'^2 - X\cdot X''}{X^4} & \frac{X'\cdot Y'}{2X^2\cdot Y^2} & \frac{X'\cdot Z'}{2X^2\cdot Z^2} \\ \frac{X'\cdot Y'}{2X^2\cdot Y^2} & \frac{Y'^2 - Y\cdot Y''}{Y^4} & \frac{Y'\cdot Z'}{2Y^2\cdot Z^2} \\ \frac{X'\cdot Z'}{2X^2\cdot Z^2} & \frac{Y'\cdot Z'}{2Y^2\cdot Z^2} & \frac{Z'^2 - Z\cdot Z''}{Z^4} \end{pmatrix} = 0 \end{cases}. \tag{353}$$

Now we demand only the traces of the two resulting equations above, which are equal apart from the sign, to become zero:

$$\text{Trace}\left(\frac{X'^2 - X \cdot X''}{2X^4}, \frac{Y'^2 - Y \cdot Y''}{2Y^4}, \frac{Z'^2 - Z \cdot Z''}{2Z^4}\right) = 0. \qquad (354)$$

For the time being, we ignore the fact that (354) is only a part of the whole picture, consider it an intrinsic degree of freedom, intend to solve it and later try to construct zero sums out of the subsequent solutions for the complete $g^{\alpha\beta}\delta_g R_{\alpha\beta}$ term within certain volumes of space. Just as the variation of the Einstein–Hilbert action allows us to do, when giving us a free choice about the surface over which we integrate the $g^{\alpha\beta}\delta_g R_{\alpha\beta}$ term.

This gives us now three equations of the form (354) for all three coordinates x, y, z, namely

$$-\partial_x^2 X \cdot X + (\partial_x X)^2 = 0$$
$$-\partial_y^2 Y \cdot Y + (\partial_y Y)^2 = 0$$
$$-\partial_z^2 Z \cdot Z + (\partial_z Z)^2 = 0, \qquad (355)$$

leading to the solutions

$$X[x] = C_x \cdot e^{\mu_x \cdot x}$$
$$Y[y] = C_y \cdot e^{\mu_y \cdot y}$$
$$Z[z] = C_z \cdot e^{\mu_z \cdot z}; \quad \mu_x, \mu_y, \mu_z \in C. \qquad (356)$$

As before, we can bring (355) in the two-equation form (338) for each of the functions X, Y, Z:

$$0 = g'' - \mu^2 \cdot g$$

$$0 = -\mu^2 \cdot g^2 + (g')^2 \Rightarrow \begin{cases} i \cdot \mu \cdot g + g' \\ i \cdot \mu \cdot g - g' \end{cases} = 0; \quad i = \sqrt{-1}$$

with $g = \{X[x], Y[y], Z[z]\}; \quad \mu = \{\mu_x, \mu_y, \mu_z\}. \qquad (357)$

Now we assume the three functions X, Y, Z to be the result of a separation approach for a function $\Psi[x, y, z]$ for a whatever partial differential equation, with

$$\Psi[x, y, z] = X[x] \cdot Y[y] \cdot Z[z]. \qquad (358)$$

Applying the time-independent Schrödinger equation* gives

$$\left[-\frac{\hbar^2}{2 \cdot m}\Delta + (V - E_n)\right]\Psi \equiv [-\lambda \cdot \Delta + (V - E_{nkl})]\,\Psi = 0$$

$$\frac{\left[-\lambda \cdot \left(\partial_x^2 X + \partial_y^2 Y + \partial_z^2 Z\right) + (V - E_{nkl})\,\Psi\right]}{\Psi} = 0$$

$$-\lambda \cdot \left(\frac{\partial_x^2 X}{X} + \frac{\partial_y^2 Y}{Y} + \frac{\partial_z^2 Z}{Z}\right) + (V - E_{nkl}) = 0. \quad (359)$$

With separable eigenvalues $E_{nkl} \rightarrow \{E_n, E_k, E_l\}$ and potentials $V = V_x[x] + V_y[y] + V_z[z]$, we obtain the three equations

$$-\lambda \cdot \partial_x^2 X + (V_x - E_n) \cdot X = 0$$

$$-\lambda \cdot \partial_y^2 Y + \left(V_y - E_k\right) \cdot Y = 0$$

$$-\lambda \cdot \partial_z^2 Z + (V_z - E_l) \cdot Z = 0, \quad (360)$$

which is exactly the same situation as we had in the one-dimensional case.

It can easily be shown that our choice of metric (347) does not compromise the zero outcome for the Einstein tensor. This condition is still fulfilled, but we leave it to the reader to prove this by a tedious but simple calculation. Thus, we only have to worry about the shear components coming with the summands $\left[g^{\sigma\kappa}\Gamma_{\alpha\beta}^{\lambda}\right]_{,\sigma}$, $\Gamma_{\sigma\rho}^{\rho}g^{\sigma\kappa}\Gamma_{\alpha\beta}^{\lambda}$ inside the $g^{\alpha\beta}\delta_g R_{\alpha\beta}$ term. There now, we intend to solve the problem, by summing up (integrating) over a sufficiently big region of space. Therefore, we need to construct a substructure allowing us piecewise settings for the solutions (356) such, that a sum over a certain region gives only constants for the bigger scale. The simplest way to obtain a solution out of (356) also fulfilling the conditions for $\left[g^{\sigma\kappa}\Gamma_{\alpha\beta}^{\lambda}\right]_{,\sigma} = \Gamma_{\sigma\rho}^{\rho}g^{\sigma\kappa}\Gamma_{\alpha\beta}^{\lambda} = 0$, however, would be to Fourier- or Laplace-integrate (sum) over enough "frequencies" μ_k with

---

*The author is aware of the fact that a comparison with the Klein–Gordon and/or Dirac equation might have been more appropriate, because the Schrödinger equation is only a non-relativistic approximation. But as the equation is so popular we will perform a first simple comparison here in order to demonstrate that quantum theory really already was inside the Einstein–Hilbert equations (only that they had ignored it). Later in the paper, we will also take care about Klein–Gordon and Dirac.

the conditions

$$F[\kappa] = F_F[\kappa] + F_L[\kappa]$$

$$F_F[\kappa] = \int_{-\infty}^{+\infty} C_\kappa^F[\mu_\kappa] \cdot e^{-i \cdot \mu_\kappa \cdot x} d\mu_\kappa \equiv const_\kappa^F$$

$$F_L[\kappa] = \int_{0}^{+\infty} C_\kappa^L[\mu_\kappa] \cdot e^{-\mu_\kappa \cdot x} d\mu_\kappa \equiv const_\kappa^L$$

$$F = \{X, Y, Z\}; \quad \kappa = \{x, y, z\}; \quad \mu_x, \mu_y, \mu_z \in R. \qquad (361)$$

We only need to make sure that the regions observed are big enough to achieve the desired constant result. This, however, is nothing but the classical expansion of the surface over which the integration has to be performed for a surface integral in order to see that it gives a zero sum. This is what Hilbert has done, but he did it without considering any intrinsic degrees of freedom of the second term $(g^{\alpha\beta} \delta_g R_{\alpha\beta})$ of his variation problem. Yes, we emphasize again: Hilbert was perfectly right in discarding of the term within the holistic picture he was obviously considering. This, however, does not mean that we should refrain from considering possible intrinsic non-zero or non-constant solutions as long as they can be made zero sums by summing them up.

In other words, the quantum effects should level each other out on bigger scales.

Thus, we have seen that on smaller scales intrinsic solutions can be extracted from discarded parts of the classical Einstein–Hilbert action also in 3D, which do not compromise the zero result for the Einstein tensor but provide a proper and rather general quantum field. By forming certain sums out of the quantum wiggle, the extended-Einstein field equations can be fulfilled on certain bigger scales. Inside smaller volumes of space, however, and depending on the frequencies at hand, a great variety of quantum systems could be constructed.

This, however, gives us something more to think about, especially when taking into consideration our observations with the one-dimensional case and the evaluation of proper distance, which suddenly resulted in an additional, imaginary dimension.

For instance, by filling the universe with harmonic oscillators in all three spatial dimensions, one might not only obtain inhabitable space but also additional imaginary coordinates, which already—somehow—might present time. In such a case, we would notice that we had obtained time out of space without actually adding it in.

As with the one-dimensional case, the space, even though mathematically already "in existence," which is to say defined and placed, even calculated with and physically manifested as a harmonic oscillator, provides no proper distances as long as no energy is added into it. Thus, space without energy is not inhabitable for anything and thus, it does not exist for anything else but itself (and a few mathematicians perhaps), even though it is there . . . somehow. The same holds for the imaginary dimension the space creates the moment it has taken on energy (meaning certain states of excitement). Interpreting these additional imaginary dimensions as time, we also have to conclude that there is no time without energy. The practical experience for an observer of socioeconomic space-times might be rather different, though, because they usually see people with loads of energy having less time than people without energy. This, however, is only the mixing up of the two expressions "there is no time" and "there is no time to waste." By substitution of the word "energy" with "brains" (with the latter also being associated with energy, at least for ordinary entities) we might now understand why so many politicians and journalists are actually a waste of both space and time.

Figure 49 shows the spatial wiggle as resulting from the discarded terms of the Einstein–Hilbert action for a 3D harmonic oscillator in a certain state of excitement. With the "wiggles" becoming shallower and thinned out for bigger distances from the center of the oscillator, space ceases to exist. We conclude that it should be possible to construct a suitable universe by positioning enough harmonic oscillators in a mathematical network and just by exciting them in the right way plus by allowing them to interact we might find a good model for our real space-time. By doing so, however, we should not forget that after forming certain quantum states like the harmonic oscillator, still a second summing up is required, which assures the overall disappearance of the $g^{\alpha\beta}\delta_g R_{\alpha\beta}$ term on bigger scales. This leads to a binding of the potentially rather distant quantum objects and one might interpret this binding as "entanglement." One might even be so bold and interpret the formation of the zero sum as something similar to the breakdown of the classical wave function.

One might also conclude here that the nature of quantum physics or any quantum appearance is just based on the fact that the universe makes use of the intrinsic degrees of freedom the curvature of space offers.

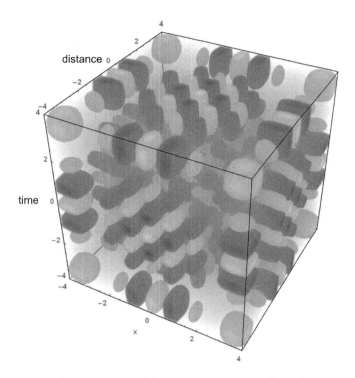

Figure 49 Metric distortion caused by a 3D harmonic oscillator for the eigenstates $n = 6$ in all three spatial directions. Subsequent evaluation of proper distances results in inhabitable space only in the vicinity of the center. For more proper space, either more oscillators or more energy needs to be added.

For completeness, one should point out that string solutions also in higher dimensions are always automatically fulfilling the extended Einstein–Hilbert action. Thus, metric approaches of the form

$$
g_{\sigma\kappa} = \left\{ \begin{pmatrix} X\,[x] & 0 & 0 \\ 0 & 1 & 0 \\ 0 & 0 & 1 \end{pmatrix}, \begin{pmatrix} 1 & 0 & 0 \\ 0 & Y\,[y] & 0 \\ 0 & 0 & 1 \end{pmatrix}, \begin{pmatrix} 1 & 0 & 0 \\ 0 & 1 & 0 \\ 0 & 0 & Z\,[z] \end{pmatrix} \right\} \tag{362}
$$

do always give an intrinsic solution of the form as shown in (356) to the coordinate bearing the functional term instead of "1," but zero if applied to (329). This is extremely practical if one intends to construct a universe out of string-like elements, especially as the statement also holds in any higher dimension.

As a natural byproduct of the intrinsic solution or quantum wiggles, and the necessary condition for the zero sum in the complete Einstein–Hilbert action, which led us to (361), we obtain the flat space solution, because all solutions summed up are constants on bigger scales. As nothing hinders us to make them complex, we might "construct" us the time we need simply out of the metric solution of our so far purely "spatial space" (more in section "The Creation of Space (and Time)").

Thus, again, we have seen, now also for the 3D case, that we can derive the metric distortion equivalent to a certain three-dimensional stationary quantum problem with (or without) potential. We have seen again that the quantum equations are part of the neglected terms of the Einstein–Hilbert action.

Nevertheless, we should also be able to find a proper connection to the technique of the intelligent zero and its auxiliary function (previous sections of this book and [9]), which we will consider in the next section.

Connection with the Technique of the "Intelligent Zero" of a Line Element

There must be a connection of the neglected terms of the Einstein–Hilbert action (331) with the method of the intelligent zero as used in our previous sections. As a very simple method for the quantization of arbitrary smooth spaces respectively space-times, we are interested in seeing how the two different approaches could be brought together. In order to make this connection visible we abbreviate and simplify as follows:

$$g^{\sigma\kappa} \to \gamma^{\sigma\kappa}; \quad \Gamma^{\lambda}_{\beta\sigma} \overset{\approx}{\longrightarrow} \frac{\gamma^{\lambda\kappa}}{2} f_{\beta\sigma,\kappa} \overset{\approx}{\longrightarrow} \frac{g}{2} f \quad \text{with} \quad f \sim \frac{g'}{g}. \tag{363}$$

Considering the variation of the Ricci tensor (327) we see terms of the form

$$\left(\frac{1}{2g} f\right)' + \frac{1}{2g} f^2 = \frac{1}{2g}\left[\frac{gf' - g'f}{g} + f^2\right] = \frac{1}{g}\left[\frac{gf'}{g} - \frac{g'}{g}f + 2 \cdot f^2\right]$$

$$= \frac{1}{g}\left[\frac{g''g - (g')^2}{g^2} - f^2 + f^2\right]$$

$$= \frac{1}{g}\left[\frac{g''g}{g^2} - f^2\right] \simeq \left[\frac{(f')^2}{\mu^2 g^2} - \frac{f^2}{g}\right]. \tag{364}$$

Thus, with an approximation

$$\mu^2 g'' \simeq (f')^2 \tag{365}$$

we can bring the connection-terms (364) we see from (327) in the form

$$\frac{(f')^2}{g^2} - \mu^2 \frac{f^2}{g} \tag{366}$$

and now exactly have the starting point of the "intelligent zero" technique where we have been able to derive the metric Dirac, Klein–Gordon and Schrödinger equations [31]. Thereby the constants μ have to be chosen such that comparing with Klein–Gordon, Dirac or Schrödinger fixes them to

$$\mu = \frac{c \cdot m}{\hbar} \tag{367}$$

with $\hbar = h/(2\pi)$ (h ... Planck's constant), m gives the mass and c denotes the speed of light in vacuum.

We see that the setting (363) is equivalent to an ansatz where the complex form of (327) is approximated by a function* f taking on the role of the connection $\Gamma^\lambda_{\beta\sigma}$ and a metric $\gamma^{\sigma\kappa}$ we assume to combine classical and quantum states. For many applications it did suffice to simply set the combined metric equal to the classical one, which is to say we have used $\gamma^{\sigma\kappa} = g^{\sigma\kappa}$. Doing so, we obtained the classical quantum theory. The auxiliary function or functional from f then was kind of equivalent to the classical wave function. It was shown in [9] how the approximation (365) could be generalized in order to reach states of lower approximation by taking higher orders of derivatives into account. Another way to derive at the "technique of the intelligent zeros" is by contraction of the extended-Einstein field equations and a proper choice of metric approaches (c.f. section "Eigenequations Derived From $\delta R_{\alpha\beta}$.")

Now we see, however, that this procedure was just an approximation of the complete solution, directly following out of the Einstein–Hilbert action if using the intrinsic degrees of freedom residing in the latter.

Nevertheless, as we will see later and as one could probably easily conclude from the complex equations resulting from (329) without the negligence of (327) in higher dimensions, we should not hastily discard of the method of the "intelligent zero," as it is a simple, straightforward, quick and even quite accurate technique of the quantization of arbitrary smooth spaces in cases of certain states of the Ricci scalar (like $R = 0$, $R = $ constant, or being of the kind $R = \text{const}/F$ [c.f. section "Examples Resulting in the Classical Quantum Theory"]).

---

*It needs to be pointed out that in general (c.f. [31] and previous sections of this book) f needs to be considered a base-vector function.

Theory: The Conjecture $\delta R_{\alpha\beta}$ = Matter & Energy and the Extended-Einstein Field Equations

At the beginning of the theory section it was hinted that by applying the technique of the intelligent zero to the line element and assuming a substructure, structures appear, which we might interpret as matter terms [27].

With the complete Einstein–Hilbert action giving the quantum theory via certain intrinsic degrees of freedom, we might expect directly to obtain also this very matter as a result of our revision of the Einstein field equations respectively the Einstein–Hilbert action. However, as we already have seen with the simple examples of the one- and the three-dimensional space, we only obtained "effective quantum zeros" if considering metrics too simple, which is to say like (333) or (347). We see this if writing down the complex equation following from a complete variation of (1) (without the "artificial" matter term). Combining the classical Einstein field equations with the variation of the Ricci tensor, which was left out by Hilbert, the complete variation does now reads

$$\delta_g W = 0 = \int_V d^n x \sqrt{-g} \left[-G^{\kappa\lambda} \delta_g g_{\kappa\lambda} + g^{\alpha\beta} \delta_g R_{\alpha\beta} \right]$$

$$= \int_V d^n x \sqrt{-g} \left[-G^{\kappa\lambda} - g^{\alpha\beta} \left(A \cdot M_{\alpha\beta}^{\lambda\kappa} - \Lambda \right) \right] \delta_g g_{\kappa\lambda}, \quad (368)$$

which gives us

$$0 = G^{\kappa\lambda} + g^{\alpha\beta} \left(A \cdot M_{\alpha\beta}^{\lambda\kappa} + \Lambda \right). \quad (369)$$

The scalar A shall take care about the fact that—for the time being—we do not know what ratio there is between the portion of $g^{\alpha\beta} \delta_g R_{\alpha\beta}$ acting as differential term in (369) or disappearing as surface integral term. The full form of the equation above would read

$$0 = R^{\kappa\lambda} - \frac{1}{2} R g^{\kappa\lambda} + \Lambda g^{\kappa\lambda} + A \cdot g^{\alpha\beta} M_{\alpha\beta}^{\lambda\kappa}. \quad (370)$$

We might name the equations above the extended-Einstein field equations or the Einstein–Hilbert field equations. Thus, the matter part of these equations seems to be

$$\text{energy \& matter} = A \cdot g^{\alpha\beta} M_{\alpha\beta}^{\lambda\kappa}$$

$$
= A \frac{g^{\alpha\beta}}{2}
\begin{pmatrix}
g^{\sigma\rho} GX_{\sigma\beta\rho}^{\kappa\lambda\eta} g_{\alpha}^{\omega} \delta g_{\kappa\lambda;\eta\omega} + g^{\omega\rho} GX_{\alpha\beta\rho}^{\kappa\lambda\eta} \delta g_{\kappa\lambda;\eta\omega} \\[6pt]
- \left[\left[g^{\sigma\rho} GX_{\sigma\beta\rho}^{\kappa\lambda\eta} \right]_{,\alpha} + g^{\sigma\rho} GX_{\sigma\beta\rho}^{\kappa\lambda\eta} \left(\Gamma_{\kappa\alpha}^{\varpi} g_{\varpi}^{\kappa} + \Gamma_{\lambda\alpha}^{\varpi} g_{\varpi}^{\lambda} + \Gamma_{\eta\alpha}^{\varpi} g_{\varpi}^{\eta} \right) \right] \delta g_{\kappa\lambda;\eta} \\[6pt]
+ \left[\left[g^{\sigma\rho} GX_{\alpha\beta\rho}^{\kappa\lambda\eta} \right]_{,\sigma} + g^{\sigma\rho} GX_{\alpha\beta\rho}^{\kappa\lambda\eta} \left(\Gamma_{\kappa\sigma}^{\varpi} g_{\varpi}^{\kappa} + \Gamma_{\lambda\sigma}^{\varpi} g_{\varpi}^{\lambda} + \Gamma_{\eta\sigma}^{\varpi} g_{\varpi}^{\eta} \right) \right] \delta g_{\kappa\lambda;\eta} \\[6pt]
+ \left[-\Gamma_{\mu\alpha}^{\sigma} g^{\mu\rho} GX_{\sigma\beta\rho}^{\kappa\lambda\eta} - \Gamma_{\beta\sigma}^{\mu} g^{\sigma\rho} GX_{\alpha\mu\rho}^{\kappa\lambda\eta} + \Gamma_{\sigma\mu}^{\mu} g^{\sigma\rho} GX_{\alpha\beta\rho}^{\kappa\lambda\eta} + \Gamma_{\alpha\beta}^{\sigma} g^{\mu\rho} GX_{\sigma\mu\rho}^{\kappa\lambda\eta} \right] \delta g_{\kappa\lambda;\eta}
\end{pmatrix}.
$$

$$(371)$$

However, as said before and explicitly evaluated in [40], there are more options, because surface- and subsurface terms form higher dimensions could couple in to our space-time leading an even more complex matter picture.

Most Symmetric and Isotropic Virtual Matter Solutions in 2D, 3D, and 4D

More or less as some kind of extended introduction, but also to demonstrate what else might be behind the idea of using the intrinsic degrees of freedom residing in the neglected terms of the Einstein–Hilbert action, we briefly want to consider very simple cases in four dimensions. A more systematic investigation will follow.

With the purely diagonal (shear-free) metric approach of

$$
g_{\sigma\kappa} =
\begin{pmatrix}
f[x] & 0 & 0 & 0 \\
0 & f[x] & 0 & 0 \\
0 & 0 & f[x] & 0 \\
0 & 0 & 0 & f[x]
\end{pmatrix}
\tag{372}
$$

Eq. (371) gives the following:

$$
\text{energy \& matter} = g^{\alpha\beta}
\begin{pmatrix}
-\left[g^{\sigma\kappa} \Gamma_{\beta\sigma}^{\lambda} \right]_{,\alpha} + \left[g^{\sigma\kappa} \Gamma_{\alpha\beta}^{\lambda} \right]_{,\sigma} \\
-\Gamma_{\sigma\alpha}^{\mu} g^{\sigma\kappa} \Gamma_{\beta\mu}^{\lambda} - \Gamma_{\beta\mu}^{\sigma} g^{\mu\kappa} \Gamma_{\sigma\alpha}^{\lambda} \\
+\Gamma_{\alpha\beta}^{\sigma} g^{\rho\kappa} \Gamma_{\sigma\rho}^{\lambda} + \Gamma_{\sigma\rho}^{\rho} g^{\sigma\kappa} \Gamma_{\alpha\beta}^{\lambda}
\end{pmatrix}
$$

$$
=
\begin{pmatrix}
\frac{f'^2 - f \cdot f''}{2 f^4} & 0 & 0 & 0 \\
0 & 3 \cdot \frac{f'^2 - f \cdot f''}{2 f^4} & 0 & 0 \\
0 & 0 & \frac{f'^2 - f \cdot f''}{2 f^4} & 0 \\
0 & 0 & 0 & \frac{f'^2 - f \cdot f''}{2 f^4}
\end{pmatrix}
= 0. \quad (373)
$$

As we explicitly want to have intrinsic or virtual solutions, we have set the result equal to zero, meaning that there is not matter. Again the general solution is

$$f[x] = C \cdot e^{\mu \cdot x}; \quad \mu \in C. \tag{374}$$

Many interesting (matter and energy) solutions can be found this way, many of which resembling particles we know.

The more interesting thing, however, is that we will not obtain such a constructive solution in the cases 2D and 3D. Here the resulting equations and solutions are for the 2D case:

$$E \& M \equiv \text{energy \& matter} = 0 = \begin{pmatrix} \frac{2f'^2 - f \cdot f''}{2f^4} & 0 \\ 0 & \frac{2f'^2 - f \cdot f''}{2f^4} \end{pmatrix} \Rightarrow f[x] = \frac{C_1}{x + C_2} \tag{375}$$

and the 3D case:

$$E \& M = 0 = \begin{pmatrix} \frac{3f'^2 - 2f \cdot f''}{2f^4} & 0 & 0 \\ 0 & \frac{3f'^2 - 2f \cdot f''}{4f^4} & 0 \\ 0 & 0 & \frac{3f'^2 - 2f \cdot f''}{4f^4} \end{pmatrix} \Rightarrow f[x] = \frac{C_1}{(x + 2C_2)^2}. \tag{376}$$

We find that only in 4D, we obtain an exponential solution.

For completeness, we also give a possible generalization. Setting the argument of the solutions (374–376),

$$\xi = g_t t + g_x x + g_y y + g_z z \tag{377}$$

(correspondingly reduced for the 2D and 3D cases), a solution can be found with

$$f[x] \Rightarrow f[\xi]. \tag{378}$$

Four Most Simple Solutions for the Whole Thing in 4D: The Matter and Antimatter Asymmetry and Why Time is Different

We use the same approach as in the sections before in the one- and the three-dimensional case:

$$g_{\sigma\kappa} = \begin{pmatrix} T[t] & 0 & 0 & 0 \\ 0 & X[x] & 0 & 0 \\ 0 & 0 & Y[y] & 0 \\ 0 & 0 & 0 & Z[z] \end{pmatrix} \tag{379}$$

and we apply it to the whole extended-Einstein field equations (370) without cosmological constant. We obtain

$$
\begin{pmatrix}
0 & \dfrac{T' \cdot X'}{4c^2 T^2 X^2} & \dfrac{T' \cdot Y'}{4c^2 T^2 Y^2} & \dfrac{T' \cdot Z'}{4c^2 T^2 Z^2} \\
\dfrac{T' \cdot X'}{4c^2 T^2 X^2} & 0 & \dfrac{X' \cdot Y'}{4X^2 Y^2} & \dfrac{X' \cdot Z'}{4X^2 Z^2} \\
\dfrac{T' \cdot Y'}{4c^2 T^2 Y^2} & \dfrac{X' \cdot Y'}{4X^2 Y^2} & 0 & \dfrac{Y' \cdot Z'}{4Y^2 Z^2} \\
\dfrac{T' \cdot Z'}{4c^2 T^2 Z^2} & \dfrac{X' \cdot Z'}{4X^2 Z^2} & \dfrac{Y' \cdot Z'}{4Y^2 Z^2} & 0
\end{pmatrix} = 0. \tag{380}
$$

Again, for brevity, we have defined

$$
T' = \partial_t T\,[t] = \frac{\partial T\,[t]}{\partial t}; X' = \partial_x X\,[x]\,; Y' = \partial_y Y\,[y]\,; Z' = \partial_z Z\,[z]\,. \tag{381}
$$

With the solutions being additive in the way $F\,[\kappa] = F_1[\kappa] + F_2[\kappa] + \ldots$ ($\kappa = t, x, y, z; F = T, X, Y, Z$), we have four possible solutions based on the construction of $F' = 0$. These four possibilities are

$$
\begin{aligned}
(a) \quad & T' = X' = Y' = 0 \\
(b) \quad & T' = X' = Z' = 0 \\
(c) \quad & T' = Y' = Z' = 0 \\
(d) \quad & X' = Y' = Z' = 0
\end{aligned} \tag{382}
$$

As it was said earlier, the non-zero terms in (380) are all resulting from the discarded part of the classical Einstein–Hilbert action, which means the term $\delta R_{\alpha\beta}$. And as this term could be made to zero in perfectly continuous space simply by choosing a proper boundary to integrate over (the Hilbert way), we here can chose to make the terms in (380) zero by summing certain solutions in suitable regions of space up. In contrast to Hilbert, we simply used the intrinsic degrees of freedom before actually summing everything up to zero. We also point out that so far, one degree of freedom was not used and this is the number of dimension. As there is no rule telling the universe in how many dimensions it has to exist and, what is more, on what scale what number of dimensions have to be "active" (not compactified) we should not expect to life in a universe only having one certain fixed number of dimensions. Also using this degree of freedom might lead to yet another possibility for solving the extended-Einstein field equations (for more, please see [32]).

We directly deduce from (382) that the universe has the option to make one dimension different from the others and we may leave it to the reader to guess which one this might be in a universe with 4 dimensions . . . especially if it concerns our universe.

We also see the option for an asymmetry of solutions with positive and negative derivatives for one of the dimensions. Knowing that the typical matter solution had the term $\text{Exp}[-i * \mu * t]$, while it is $\text{Exp}[+i * \mu * t]$ for antimatter, we also might have found a way to explain the possibility of a surplus of matter in bigger regions of the universe. Interestingly, this would be option d) in our universal selection of possibilities as given in (382) and it would come with the condition of X'', which is—what a surprise—a perfectly flat space on bigger scales.

The Two-Dimensional Case

It is a straightforward evaluation to demonstrate that a metric approach of the form

$$g_{\alpha\beta} = \begin{pmatrix} X[x] & 0 \\ 0 & Y[y] \end{pmatrix} \tag{383}$$

applied to an intrinsic degree of freedom of (327) as done before with the 1D and 3D case, leading to equations of the form (331), would give us general solutions of the following kind again:

$$X[x] = C_x \cdot e^{\mu_x \cdot x}$$
$$Y[y] = C_y \cdot e^{\mu_y \cdot y}; \quad \mu_x, \mu_y \in C. \tag{384}$$

Applied to the harmonic oscillator we result in solutions as already developed for the 1D case and subsequently, when evaluating the proper distance, we also obtain an additional dimension. Meaning, a two-dimensional space consisting of oscillators and giving some energy automatically evolves into dimensions. This sums up to our space time with an underlying two-dimensional subspace . . . potentially a concept being quite attractive for all supporters of the holographic principle [33–35].

Intermediate Result: The *n*-Dimensional Case

In every number of dimensions approaches of the form (372), which in a *n*-dimensional case would read

$$g_{\alpha\beta} = \begin{pmatrix} f[x_k] & \cdots & 0 \\ \vdots & \ddots & \vdots \\ 0 & \cdots & f[x_k] \end{pmatrix}, \tag{385}$$

can give solutions of the kind

$$F_k[x_k] = C_{x_k} \cdot e^{\mu_{x_k} \cdot x_k}; \quad \mu_{x_k} \in C \tag{386}$$

if being considered in connection with the traditionally neglected parts of the Einstein–Hilbert action in using the intrinsic degrees of freedom residing within the latter. We interpret these solutions as the virtual quantum background field. If applied to the whole variation term of the Ricci tensor and demanding this to be zero, we obtain rather different solutions for different numbers of dimensions (c.f. (374), (375) and (376) for 4D, 2D, and 3D, respectively).

We also find that approaches of the form (347), (379), (383), if applied to the complete extended-Einstein field equations, are also giving quantum-like equations and subsequent quantum solutions of the form

$$F_0[x_0] = C_{x_0} \cdot e^{\mu_{x_0} \cdot x_0}; \quad \mu_{x_0} \in C$$
$$\vdots$$
$$F_k[x_k] = C_{x_k} \cdot e^{\mu_{x_k} \cdot x_k}; \quad \mu_{x_k} \in C \tag{387}$$
$$\vdots$$
$$F_D[x_D] = C_{x_D} \cdot e^{\mu_{x_D} \cdot x_D}; \quad \mu_{x_D} \in C; \quad F_k = X, Y, Z, \ldots$$

with the additional boundary condition of $F_k'[x_k] = 0$ for at least $D-1$ of the D functions F_k. As these functions are additive, not only the boundary conditions above can be satisfied, but also a great variety of space-time constructions, like a space made out of harmonic oscillators, seem to be possible.

Antimatter and Spin

It can be shown that a metric approach of the form

$$g_{\sigma\kappa} = \begin{pmatrix} F[t,x,y,z] & 0 & 0 & 0 \\ 0 & F[t,x,y,z] & 0 & 0 \\ 0 & 0 & F[t,x,y,z] & 0 \\ 0 & 0 & 0 & F[t,x,y,z] \end{pmatrix} \tag{388}$$

set into the extended-Einstein field equations (370) (with the cosmological constant set to zero) can be solved by a general function F with

$$F[t,x,y,z] = F[\xi] = F[g_t t + g_x x + g_y y + g_z z] = \frac{C_2}{(\xi + C_1)^4} \tag{389}$$

and the additional condition

$$g_t^2 + g_x^2 + g_y^2 + g_z^2 = 0. \tag{390}$$

Interestingly, we find that there are always two options for each coordinate. If we demand that (390) should be fulfilled also for those cases where the dependency with respect to a certain coordinate vanishes, we should demand

$$
\begin{aligned}
g_t^2 + g_x^2 + g_y^2 + g_z^2 = 0 &\quad\Rightarrow\quad g_t = \pm i \cdot \sqrt{g_x^2 + g_y^2 + g_z^2} \\
g_x^2 + g_y^2 + g_z^2 = 0 &\quad\Rightarrow\quad g_x = \pm i \cdot \sqrt{g_y^2 + g_z^2} \\
g_y^2 + g_z^2 = 0 &\quad\Rightarrow\quad g_y = \pm i \cdot g_z
\end{aligned}
\tag{391}
$$

We might interpret the various options regarding the parameters g_t, g_x, g_y as antimatter and matter, spin-up and spin-down, and two settings of iso-spin, respectively.

A metric disturbance of the form (389) can be matter or antimatter if having access to all 4 dimensions of space-time. It evolves a 2-option spin and has to decide for one of the two spin options if being forced to become stationary (no time dependency). The moment the metric disturbance will also be restricted to only two dimensions, it must sacrifice its last degree of freedom and also fixes the iso-spin to one of two options. We recognize the typical quantum theoretical restriction that not all these properties could be measured in one go, but require special conditions for each. We also realize that the disturbance (389) or a similar solution, if being restricted to a brane (two-dimensional manifold), like the $z = 0$ plane

$$
\begin{aligned}
g_t^2 + g_x^2 + g_y^2 = 0 &\quad\Rightarrow\quad g_t = \pm i \cdot \sqrt{g_x^2 + g_y^2} \\
g_x^2 + g_y^2 = 0 &\quad\Rightarrow\quad g_x = \pm i \cdot g_y
\end{aligned}
\tag{392}
$$

has only two parameter options left, namely the matter–antimatter option and spin. If being restricted to a one-dimensional subspace, $z = y = 0$

$$g_t^2 + g_x^2 = 0 \quad\Rightarrow\quad g_t = \pm i \cdot g_x, \tag{393}$$

the disturbance could only become either matter or antimatter, as all other options are "frozen out."

An Adapted Schwarzschild Solution

It was found that a most simple function of the form (389) gives a solution for the extended-Einstein field equations (370). We assume that there must be some connection with the Schwarzschild solution [36] and consider its isotropic form

$$\{g_{00}, g_{11}, g_{22}, g_{33}\}$$

$$= \left\{ \frac{\left(1 - \frac{r_s}{4 \cdot \varrho}\right)^2}{\left(1 + \frac{r_s}{4 \cdot \varrho}\right)^2}, \left(1 + \frac{r_s}{4 \cdot \varrho}\right)^4, \left(1 + \frac{r_s}{4 \cdot \varrho}\right)^4, \left(1 + \frac{r_s}{4 \cdot \varrho}\right)^4 \right\}. \quad (394)$$

Now we use a simple form of our solution (389) with only two coordinate-dependencies like

$$F\left[\tilde{\xi}\right] = \frac{C_2}{\left(\tilde{\xi} + C_1\right)^4} = \frac{C_2}{\left(g_x \cdot \xi + C_1\right)^4} = \frac{C_2}{\left(g_x \left(x \pm i \cdot \rho\right) + C_1\right)^4}; \quad \rho = c \cdot t \vee y \vee z$$

$$(395)$$

and set the spatial components of both solutions to be equal. This gives us the following relation for our coordinate ξ:

$$\xi = -\frac{1}{g_x} \cdot \frac{C_1 \cdot r_s + 4 \cdot C_1 \cdot \varrho - 4 \cdot C_2 \cdot \varrho}{r_s + 4 \cdot \varrho}. \quad (396)$$

Series expansion of $\left(1 + \frac{r_s}{4 \cdot \varrho}\right)^4$ and $\frac{C_2}{\left(g_x \cdot \xi[\varrho] + C_1\right)^4}$ at $\varrho = \infty$ shows that only C_2 needs to be fixed to $C_2 = 1$, in order to asymptotically obtain the classical law of gravity for bigger ϱ, while g_x and C_1 are arbitrary. Thus, for purely real ξ, naturally, we obtain the classical dependency with respect to the isotropic Schwarzschild coordinate ϱ, including singularities, etc. However, as we have learned, the solution for the complete or extended-Einstein field equations is not (394), but (395) (or its more complex form (389)) and therefore need to realize that ξ could well be complex. Simply setting it $\xi \rightarrow \xi[\varrho] + i * c * t$, with both ϱ and t assumed to be real, would clearly only allow for a singularity solution at $t = 0$. Figure 50 gives an idea about the metric behavior for both solutions, the classical Schwarzschild and the adapted one in dependency on the isotropic Schwarzschild radius coordinate ϱ.

We should note that we could adjust g_x such that it resembles the typical Dirac factor for the time coordinate, which usually is $\pm i \cdot \frac{E}{\hbar} = \pm i \cdot \frac{m \cdot c^2}{\hbar}$. Applying the well-known equation for the Schwarzschild radius $r_s = \frac{2 \cdot m \cdot G}{c^2}$, we may set ($G$ . . . Newton's or gravitational constant with $G = 6.6742 \times 10^{-11}$ m^3/(kg*s^2)):

$$g_x = \frac{r_s \cdot c^3}{2 \cdot \hbar \cdot G}. \quad (397)$$

It appears as no big surprise that this gives us the parameter g_x as a fraction of the Schwarzschild radius r_s, measured in units of Planck length

$l_P = \sqrt{\frac{\hbar \cdot G}{c^3}} \approx 1.616229 \cdot 10^{-35} m$ and divided by the latter. This way we would even have obtained a matter $-i \cdot \frac{E}{\hbar} \cdot t = -i \cdot \frac{m \cdot c^2}{\hbar} \cdot t$ and an antimatter $+i \cdot \frac{E}{\hbar} \cdot t = +i \cdot \frac{m \cdot c^2}{\hbar} \cdot t$ Schwarzschild object. This means that the surface area obtained by the product of our coordinate x (c.f. Eq. (395), term after last "=") times r_s is measured in units of half Planck area. With measuring the coordinates in such a way we might also suspect the principle area uncertainty to be connected with the parameters given in (397). When evaluating a certain stretch of proper distance along the x coordinate, it has to be done via

$$\Delta s_x = \int_{x_0}^{\Delta x + x_0} \frac{\sqrt{C_2}}{(g_x (x \pm i \cdot \rho) + C_1)^2} dx; \quad \rho = c \cdot t \vee y \vee z$$

$$= \frac{\Delta x \cdot \sqrt{C_2}}{(C_1 + g_x (i \cdot \rho + x_0)) (C_1 + g_x (i \cdot \rho - x_0 + \Delta x))}. \quad (398)$$

We see from Fig. 50 that with increasing time the Schwarzschild object smears out or one might say perhaps "it evaporates."

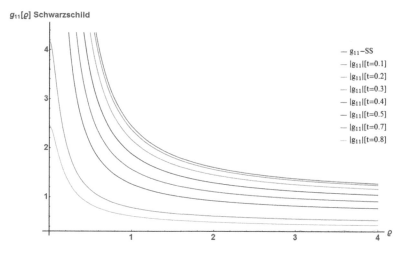

Figure 50 Spatial metric component for the adapted Schwarzschild solution compatible with the extended-Einstein field equations in comparison with the classical isotropic solution g_{11}–SS. Here we used $\xi \to \xi [\varrho] + i * c * t$. For better illustration we kept the constants in all cases equal to one ($c = C_1 = C_2 = g_x = r_s = 1$).

If we insist on constructing a stable Schwarzschild object (stable in time), we might like to define a setup like

$$F\left[\tilde{\xi}\right] = \frac{C_2}{\left(\tilde{\xi} + C_1\right)^4} = \frac{C_2}{\left(g_x \cdot \xi + C_1\right)^4} = \frac{C_2}{\left(g_x\left(x \pm i \cdot \frac{\rho}{\sqrt{2}}\right) + C_1\right)^4}; \quad \rho = y+z.$$

(399)

We see that we have two spin-like states as possible time-stable solutions resulting from the extended-Einstein field equations (370). In accordance with our considerations with respect to our g_x for the case $\rho = c * t$ we should now expect to have the same dependency, giving us

$$F\left[\tilde{\xi}\right] = \frac{C_2}{\left(\frac{r_s \cdot c^3}{2 \cdot \hbar \cdot G}\left(x \pm i \cdot \frac{y+z}{\sqrt{2}}\right) + C_1\right)^4} \equiv \frac{C_2}{\left(\frac{r_s \cdot c^3}{2 \cdot \hbar \cdot G}\left(x + i \cdot \left(\varepsilon \pm \frac{y+z}{\sqrt{2}}\right)\right)\right)^4},$$

(400)

where we have chosen an imaginary C_1 in order to assure that there is no singularity at $x = y = z = 0$ and for structural convenience, we also connected it with the factor g_x.

Eigenequations Derived from $\delta R_{\alpha\beta}$ for Shear-Free Metrics

Now we want to investigate separation approaches in various dimensions. Thereby we seek for solutions for the $\delta R_{\alpha\beta}$ only, meaning solutions of the equation (371) set to either zero or an eigenvalue form, like

$$\widehat{g}_n^{\kappa\lambda} = g^{\alpha\beta}\begin{pmatrix} -\left[g^{\sigma\kappa}\Gamma^\lambda_{\beta\sigma}\right]_{,\alpha} + \left[g^{\sigma\kappa}\Gamma^\lambda_{\alpha\beta}\right]_{,\sigma} \\ -\Gamma^\mu_{\sigma\alpha}g^{\sigma\kappa}\Gamma^\lambda_{\beta\mu} - \Gamma^\sigma_{\beta\mu}g^{\mu\kappa}\Gamma^\lambda_{\sigma\alpha} \\ +\Gamma^\sigma_{\alpha\beta}g^{\rho\kappa}\Gamma^\lambda_{\sigma\rho} + \Gamma^\rho_{\sigma\rho}g^{\sigma\kappa}\Gamma^\lambda_{\alpha\beta} \end{pmatrix}.$$

(401)

We note that a contraction like

$$\widehat{G}_n = \widehat{g}_{n|\kappa\lambda}\widehat{g}_n^{\kappa\lambda} = \widehat{g}_{n|\kappa\lambda}\left[g^{\alpha\beta}\begin{pmatrix} -\left[g^{\sigma\kappa}\Gamma^\lambda_{\beta\sigma}\right]_{,\alpha} + \left[g^{\sigma\kappa}\Gamma^\lambda_{\alpha\beta}\right]_{,\sigma} \\ -\Gamma^\mu_{\sigma\alpha}g^{\sigma\kappa}\Gamma^\lambda_{\beta\mu} - \Gamma^\sigma_{\beta\mu}g^{\mu\kappa}\Gamma^\lambda_{\sigma\alpha} \\ +\Gamma^\sigma_{\alpha\beta}g^{\rho\kappa}\Gamma^\lambda_{\sigma\rho} + \Gamma^\rho_{\sigma\rho}g^{\sigma\kappa}\Gamma^\lambda_{\alpha\beta} \end{pmatrix}\right]$$

(402)

and an expansion like

$$\widehat{G}_{n|\alpha\beta}^{\kappa\lambda} = \left(-\left[g^{\sigma\kappa}\Gamma^\lambda_{\beta\sigma}\right]_{,\alpha} + \left[g^{\sigma\kappa}\Gamma^\lambda_{\alpha\beta}\right]_{,\sigma} - \Gamma^\mu_{\sigma\alpha}g^{\sigma\kappa}\Gamma^\lambda_{\beta\mu}\right.$$
$$\left. -\Gamma^\sigma_{\beta\mu}g^{\mu\kappa}\Gamma^\lambda_{\sigma\alpha} + \Gamma^\sigma_{\alpha\beta}g^{\rho\kappa}\Gamma^\lambda_{\sigma\rho} + \Gamma^\rho_{\sigma\rho}g^{\sigma\kappa}\Gamma^\lambda_{\alpha\beta}\right)$$

(403)

is possible. Please note that so far the terms $\widehat{G}_n$ are still functions of the coordinates and no true eigenvalues.

We see that there is a huge variety of eigenvalues, eigenmetrics, and four-order eigentensors.

In Four Dimensions

We start in four dimensions and apply (388) again, but this time we incorporate the speed of light c and use the sign convention $(-, +, +, +)$:

$$
g_{\sigma\kappa} = \begin{pmatrix}
-c^2 \cdot F\,[t, x, y, z] & 0 & 0 & 0 \\
0 & F\,[t, x, y, z] & 0 & 0 \\
0 & 0 & F\,[t, x, y, z] & 0 \\
0 & 0 & 0 & F\,[t, x, y, z]
\end{pmatrix}.
$$

$$(404)$$

The subsequent eigenvalue equation (402) becomes

$$
\widehat{G}_n = \widehat{g}_{n|\kappa\lambda}\widehat{g}_n^{\kappa\lambda} = \widehat{g}_{n|\kappa\lambda}\left[g^{\alpha\beta}\begin{pmatrix} -\left[g^{\sigma\kappa}\Gamma^\lambda_{\beta\sigma}\right]_{,\alpha} + \left[g^{\sigma\kappa}\Gamma^\lambda_{\alpha\beta}\right]_{,\sigma} \\ -\Gamma^\mu_{\sigma\alpha}g^{\sigma\kappa}\Gamma^\lambda_{\beta\mu} - \Gamma^\sigma_{\beta\mu}g^{\mu\kappa}\Gamma^\lambda_{\sigma\alpha} \\ +\Gamma^\sigma_{\alpha\beta}g^{\rho\kappa}\Gamma^\lambda_{\sigma\rho} + \Gamma^\rho_{\sigma\rho}g^{\sigma\kappa}\Gamma^\lambda_{\alpha\beta} \end{pmatrix} \right]
$$

$$
= \frac{3}{F^3}\cdot\left(\frac{F\cdot\partial_t^2 F - (\partial_t F)^2}{c^2} + (\partial_x F)^2 + (\partial_y F)^2 \right.
$$

$$
\left. + (\partial_z F)^2 - F\cdot\left(\partial_x^2 F + \partial_y^2 F + \partial_z^2 F\right) \right). \tag{405}
$$

Now we define the new (true) eigenvalues $\tilde{G}_n \equiv F\cdot\widehat{G}_n$ and obtain

$$
\tilde{G}_n = \frac{3}{F^2}\cdot\left(\frac{F\cdot\partial_t^2 F - (\partial_t F)^2}{c^2} + (\partial_x F)^2 + (\partial_y F)^2 + (\partial_z F)^2 \right.
$$

$$
\left. - F\cdot\left(\partial_x^2 F + \partial_y^2 F + \partial_z^2 F\right) \right). \tag{406}
$$

Seeking a solution in the following form

$$
\begin{array}{ll}
\text{(I)} & \mu^2\cdot F = \frac{\partial_t^2 F}{c^2} - \partial_x^2 F - \partial_y^2 F - \partial_z^2 F \\
\text{(II)} & F^2\cdot\left(\frac{\tilde{G}_n}{3} - \mu^2\right) = \left(\frac{-(\partial_t F)^2}{c^2} + (\partial_x F)^2 + (\partial_y F)^2 + (\partial_z F)^2\right),
\end{array} \tag{407}
$$

and comparing the equation (I) in (407) above with the classical Klein–Gordon equation

$$
\left[-\frac{\partial_t^2}{c^2} + \Delta - \frac{m^2 c^2}{\hbar^2} \right]\Psi_{KG} = 0, \tag{408}
$$

we find the relation

$$\left[-\frac{\partial_t^2}{c^2} + \Delta - \frac{m^2 c^2}{\hbar^2}\right] \Psi_{KG} = 0 \Leftrightarrow 0 = \left[-\frac{\partial_t^2}{c^2} + \Delta_{x,y,z} + \mu^2\right] F, \tag{409}$$

which is already giving us the perfect match for the Klein–Gordon equation. We clearly see that

$$-\frac{m^2 c^2}{\hbar^2} = \mu^2 \Leftrightarrow \mu = i \cdot \frac{m \cdot c}{\hbar}. \tag{410}$$

We have to be careful with the Laplace operators in (409). While on the left-hand side the operator Δ truly stands for the Laplace operator, we need to emphasize that on the right-hand side $\Delta_{x,y,z}$ stands for

$$\Delta_{x,y,z} = \partial_x^2 F + \partial_y^2 F + \partial_z^2 F. \tag{411}$$

If we insist on a complete reproduction of the Klein–Gordon equation, we have to adjust our metric approach as follows:

$$g_{\sigma\kappa} = \begin{pmatrix} -c^2 \cdot \Psi_{KG} & 0 & 0 & 0 \\ 0 & \tilde{g}_{11} \cdot \Psi_{KG} & 0 & 0 \\ 0 & 0 & \tilde{g}_{22} \cdot \Psi_{KG} & 0 \\ 0 & 0 & 0 & \tilde{g}_{33} \cdot \Psi_{KG} \end{pmatrix}, \tag{412}$$

where the components $\tilde{g}_{ii}$ denote the coordinate system in which the Klein–Gordon equation shall be constructed. The relationship can also be obtained for metrics with shear components, but as we are aiming for the complete solution and not a certain "Klein–Gordon part" of it, we are not going to elaborate this here. We only point out that in the same way, namely by adjusting the metric approach, one could also arrive at the "intelligent zero" technique as is was widely used by the present author before (see previous sections and references).

If applying the vectorial root extraction as given in [9, 20], we can also bring equation (II) in (407) in a form where we are able to recognize the Dirac forms. So we have

$$F^2 \cdot \left(\frac{\tilde{G}_n}{3} - \mu^2\right) \equiv -F^2 \cdot \lambda^2 = \left(\frac{-(\partial_t F)^2}{c^2} + (\partial_x F)^2 + (\partial_y F)^2 + (\partial_z F)^2\right)$$

$$\Rightarrow 0 = \begin{cases} \lambda \cdot F + i \cdot \frac{\partial_t F}{c} + \partial_x F + \partial_y F + \partial_z F, \ \lambda \cdot F + i \cdot \frac{\partial_t F}{c} + \partial_x F + \partial_y F - \partial_z F, \\ \lambda \cdot F + i \cdot \frac{\partial_t F}{c} + \partial_x F - \partial_y F + \partial_z F, \ \lambda \cdot F + i \cdot \frac{\partial_t F}{c} + \partial_x F - \partial_y F - \partial_z F, \\ \lambda \cdot F + i \cdot \frac{\partial_t F}{c} - \partial_x F + \partial_y F + \partial_z F, \ \lambda \cdot F + i \cdot \frac{\partial_t F}{c} - \partial_x F + \partial_y F - \partial_z F, \\ \lambda \cdot F + i \cdot \frac{\partial_t F}{c} - \partial_x F - \partial_y F + \partial_z F, \ \lambda \cdot F + i \cdot \frac{\partial_t F}{c} - \partial_x F - \partial_y F - \partial_z F, \\ \lambda \cdot F - i \cdot \frac{\partial_t F}{c} + \partial_x F + \partial_y F + \partial_z F, \ \lambda \cdot F - i \cdot \frac{\partial_t F}{c} + \partial_x F + \partial_y F - \partial_z F, \\ \lambda \cdot F - i \cdot \frac{\partial_t F}{c} + \partial_x F - \partial_y F + \partial_z F, \ \lambda \cdot F - i \cdot \frac{\partial_t F}{c} + \partial_x F - \partial_y F - \partial_z F, \\ \lambda \cdot F - i \cdot \frac{\partial_t F}{c} - \partial_x F + \partial_y F + \partial_z F, \ \lambda \cdot F - i \cdot \frac{\partial_t F}{c} - \partial_x F + \partial_y F - \partial_z F, \\ \lambda \cdot F - i \cdot \frac{\partial_t F}{c} - \partial_x F - \partial_y F + \partial_z F, \ \lambda \cdot F - i \cdot \frac{\partial_t F}{c} - \partial_x F - \partial_y F - \partial_z F \end{cases} . \tag{413}$$

It needs to be pointed out that not every solution of the equations (407) is also solving (406). Thus, we need to treat this kind of separation with great care and we here only made use of it in order to show the connection to the classical quantum theory. We will see below that a closed form solution of the governing equation (406) can be obtained by the means of separation approaches (see next section).

In Three Dimensions

In three dimensions the metric approach

$$g_{\sigma\kappa} = \begin{pmatrix} F\,[x,\,y,\,z] & 0 & 0 \\ 0 & F\,[x,\,y,\,z] & 0 \\ 0 & 0 & F\,[x,\,y,\,z] \end{pmatrix} \tag{414}$$

set into (402) gives

$$
\hat{G}_n = \hat{g}_{n|\kappa\lambda}\hat{g}_n^{\kappa\lambda} = \hat{g}_{n|\kappa\lambda} \left[g^{\alpha\beta} \begin{pmatrix} -\left[g^{\sigma\kappa}\Gamma^\lambda_{\beta\sigma}\right]_{,\alpha} + \left[g^{\sigma\kappa}\Gamma^\lambda_{\alpha\beta}\right]_{,\sigma} \\ -\Gamma^\mu_{\sigma\alpha}g^{\sigma\kappa}\Gamma^\lambda_{\beta\mu} - \Gamma^\sigma_{\beta\mu}g^{\mu\kappa}\Gamma^\lambda_{\sigma\alpha} \\ +\Gamma^\sigma_{\alpha\beta}g^{\rho\kappa}\Gamma^\lambda_{\sigma\rho} + \Gamma^\rho_{\sigma\rho}g^{\sigma\kappa}\Gamma^\lambda_{\alpha\beta} \end{pmatrix} \right]
$$

$$
= \frac{3\cdot\left((\partial_x F)^2 + (\partial_y F)^2 + (\partial_z F)^2\right) - 2\cdot F\cdot\left(\partial_x^2 F + \partial_y^2 F + \partial_z^2 F\right)}{F^3}. \tag{415}
$$

Again, we define the new (true) eigenvalues $\tilde{G}_n \equiv F\cdot\hat{G}_n$ and obtain

$$
\tilde{G}_n = \frac{3\cdot\left((\partial_x F)^2 + (\partial_y F)^2 + (\partial_z F)^2\right) - 2\cdot F\cdot\left(\partial_x^2 F + \partial_y^2 F + \partial_z^2 F\right)}{F^2}. \tag{416}
$$

Seeking a solution in the following form

$$
\begin{array}{ll}
\text{(I)} & \mu^2\cdot F = -\partial_x^2 F - \partial_y^2 F - \partial_z^2 F \\
\text{(II)} & F^2\cdot\frac{(\tilde{G}_n - 2\cdot\mu^2)}{3} = \left((\partial_x F)^2 + (\partial_y F)^2 + (\partial_z F)^2\right),
\end{array} \tag{417}
$$

and comparing the equation (I) in (407) above with the classical time-independent Klein–Gordon equation

$$
\left[\Delta - \frac{m^2 c^2}{\hbar^2}\right]\Psi_{KG} = 0, \tag{418}
$$

we find the relation

$$
\left[\Delta - \frac{m^2 c^2}{\hbar^2}\right]\Psi_{KG} = 0 \Leftrightarrow 0 = \left[\Delta_{x,y,z} + \mu^2\right] F, \tag{419}
$$

with the rest following just as it has been shown in 4D. For instance, the Dirac form (part II) in (417)) now becomes

$$F^2 \cdot \frac{\left(\tilde{G}_n - 2 \cdot \mu^2\right)}{3} \equiv -F^2 \cdot \lambda^2 = (\partial_x F)^2 + (\partial_y F)^2 + (\partial_z F)^2$$

$$\Rightarrow 0 = \begin{cases} \lambda \cdot F + \partial_x F + \partial_y F + \partial_z F, \ \lambda \cdot F + \partial_x F + \partial_y F - \partial_z F, \\ \lambda \cdot F + \partial_x F - \partial_y F + \partial_z F, \ \lambda \cdot F + \partial_x F - \partial_y F - \partial_z F, \\ \lambda \cdot F - \partial_x F + \partial_y F + \partial_z F, \ \lambda \cdot F - \partial_x F + \partial_y F - \partial_z F, \\ \lambda \cdot F - \partial_x F - \partial_y F + \partial_z F, \ \lambda \cdot F - \partial_x F - \partial_y F - \partial_z F, \end{cases} \cdot \quad (420)$$

As already hinted in the 4D case, if interested in perfectly reproducing the Klein–Gordon and the Dirac form, we need to adjust the metric approach (414) in such a way, that the pseudo-Laplace operator $\Delta_{x,y,z}$ in (419) becomes a differential operator acting in the coordinate system we want to have our Klein–Gordon equation in. May this be a coordinate system of the form $\tilde{g}_{\alpha\beta} = \text{Trace}\{\tilde{g}_{11}, \tilde{g}_{22}, \tilde{g}_{33}\}$, then we should use the ansatz

$$g_{\sigma\kappa} = \begin{pmatrix} \tilde{g}_{11} \cdot \Psi[x, y, z] & 0 & 0 \\ 0 & \tilde{g}_{22} \cdot \Psi[x, y, z] & 0 \\ 0 & 0 & \tilde{g}_{33} \cdot \Psi[x, y, z] \end{pmatrix}. \quad (421)$$

In Two Dimensions

In two dimensions things are getting relatively simple, because a metric approach of the form

$$g_{\alpha\beta} = \begin{pmatrix} F[x, y] & 0 \\ 0 & F[x, y] \end{pmatrix}, \quad (422)$$

set into (401) gives us the equation

$$\tilde{g}_n^{\kappa\lambda} = g^{\alpha\beta} \begin{pmatrix} -\left[g^{\sigma\kappa}\Gamma^\lambda_{\beta\sigma}\right]_{,\alpha} + \left[g^{\sigma\kappa}\Gamma^\lambda_{\alpha\beta}\right]_{,\sigma} \\ -\Gamma^\mu_{\sigma\alpha}g^{\sigma\kappa}\Gamma^\lambda_{\beta\mu} - \Gamma^\sigma_{\beta\mu}g^{\mu\kappa}\Gamma^\lambda_{\sigma\alpha} \\ +\Gamma^\sigma_{\alpha\beta}g^{\rho\kappa}\Gamma^\lambda_{\sigma\rho} + \Gamma^\rho_{\sigma\rho}g^{\sigma\kappa}\Gamma^\lambda_{\alpha\beta} \end{pmatrix} = \begin{pmatrix} \hat{E}_{2D}(F[x, y]) & 0 \\ 0 & \hat{E}_{2D}(F[x, y]) \end{pmatrix}$$

$$\hat{E}_{2D}(F[x, y]) = \hat{E}_{2D}(F) = \frac{-2 \cdot \left((\partial_x F)^2 + (\partial_y F)^2\right) + F \cdot (\partial_x^2 F + \partial_y^2 F)}{2 \cdot F^4} \Rightarrow 0.$$

$$(423)$$

With the Einstein tensor always identical zero in two dimensions, we have to set the above equation zero, too. The subsequent equation then apparently reads

$$0 = F \cdot (\partial_x^2 F + \partial_y^2 F) - 2 \cdot \left((\partial_x F)^2 + (\partial_y F)^2\right), \quad (424)$$

but we should not rule out "intelligent zeros" of the form

$$\hat{g}_n^{\kappa\lambda} - \hat{g}_n^{\kappa\lambda} \equiv \frac{\tilde{g}_n^{\kappa\lambda} - \tilde{g}_n^{\kappa\lambda}}{2 \cdot F^2} = \frac{-2 \cdot \left((\partial_x F)^2 + (\partial_y F)^2 \right) + F \cdot \left(\partial_x^2 F + \partial_y^2 F \right)}{2 \cdot F^4}$$

$$\Rightarrow F^2 \cdot \tilde{g}_n^{\kappa\lambda} - F^2 \cdot \tilde{g}_n^{\kappa\lambda} = 2 \cdot \left((\partial_x F)^2 + (\partial_y F)^2 \right) - F \cdot \left(\partial_x^2 F + \partial_y^2 F \right),$$

$$(425)$$

which is nothing but a separation approach (see more below) and again directly leads us to the classical quantum equations, if using the techniques as described above. Thus, we obtain

$$\begin{array}{ll} \text{(I)} & \mu^2 \cdot F = -\partial_x^2 F - \partial_y^2 F \\ \text{(II)} & F^2 \cdot \frac{\left(\tilde{g}_n^{\kappa\lambda} - \tilde{g}_n^{\kappa\lambda} - \mu^2 \right)}{2} = \left((\partial_x F)^2 + (\partial_y F)^2 \right), \end{array} \qquad (426)$$

and comparing the equation (I) with the classical time-independent Klein–Gordon equation

$$\left[\Delta_{2D} - \frac{m^2 c^2}{\hbar^2} \right] \Psi_{KG} = 0, \qquad (427)$$

we find the relation

$$\left[\Delta_{2D} - \frac{m^2 c^2}{\hbar^2} \right] \Psi_{KG} = 0 \Leftrightarrow 0 = \left[\Delta_{x,y} + \mu^2 \right] F. \qquad (428)$$

The Dirac form (part II) in (426)) now becomes rather simple:

$$F^2 \cdot \frac{\left(\tilde{g}_n^{\kappa\lambda} - \tilde{g}_n^{\kappa\lambda} - \mu^2 \right)}{2} \equiv -F^2 \cdot \lambda^2 = (\partial_x F)^2 + (\partial_y F)^2$$

$$\Rightarrow 0 = \left\{ \begin{array}{l} \lambda \cdot F + \partial_x F + \partial_y F, \lambda \cdot F + \partial_x F - \partial_y F, \\ \lambda \cdot F - \partial_x F + \partial_y F, \lambda \cdot F - \partial_x F - \partial_y F \end{array} \right\}. \qquad (429)$$

Again we hint that a perfect reproduction of the Klein–Gordon and the Dirac equation in the classical form requires a suitable readjustment of the metric approach (422) in such a way, that the pseudo-Laplace operator $\Delta_{x,y}$ in (428) becomes a differential operator acting in the coordinate system we want to have our Klein–Gordon equation in. May this be a coordinate system of the form $\tilde{g}_{\alpha\beta} = \text{Trace}\{\tilde{g}_{11}, \tilde{g}_{22}\}$, then we should use the ansatz

$$g_{\alpha\beta} = \begin{pmatrix} \tilde{g}_{11} \cdot \Psi[x, y] & 0 \\ 0 & \tilde{g}_{22} \cdot \Psi[x, y] \end{pmatrix}. \qquad (430)$$

We explicitly note that every holomorphic function $F[x, y] = F[x + i \cdot y]$ automatically satisfies the homogeneous equation (424).

Summing Up This Section

Here we have been able to show, that classical quantum equations can be obtained from the discarded part of the Einstein Hilbert action, which is to say the term $\delta R_{\alpha\beta}$, by a simple contraction of the resulting variation (a tensor) with the metric.

Separation Approaches

In the following we intend to apply separation approaches to the eigenvalue, eigentensor and eigen-matrix equations derived in section "Eigenequations Derived From $\delta R_{\alpha\beta}$."

The 2D Case

As in the 2D case, the eigentensor is always zero and we have the interesting fact that solutions for (401) should always give zero in two dimensions.

For completeness, we still want to solve the 2D case with an approach of the form

$$g_{\alpha\beta} = \begin{pmatrix} F[x,y] & 0 \\ 0 & F[x,y] \end{pmatrix}, \tag{431}$$

which gives us, as we already saw, the equation

$$0 = \begin{pmatrix} \hat{E}_{2D}(F[x,y]) & 0 \\ 0 & \hat{E}_{2D}(F[x,y]) \end{pmatrix}$$

$$\hat{E}_{2D}(F[x,y]) = \hat{E}_{2D}(F) = \frac{-2 \cdot \left((\partial_x F)^2 + (\partial_y F)^2 \right) + F \cdot (\partial_x^2 F + \partial_y^2 F)}{2 \cdot F^4}. \tag{432}$$

We start with the following separation approach

$$F[x,y] = X[x] \cdot Y[y]. \tag{433}$$

Setting this into (370), respectively its subsequent 2D form (432), we obtain

$$0 = \left(\frac{\partial_x^2 X}{X} + \frac{\partial_y^2 Y}{Y} \right) - 2 \cdot \left(\left(\frac{\partial_x X}{X} \right)^2 + \left(\frac{\partial_y Y}{Y} \right)^2 \right)$$

$$\Rightarrow -\frac{\partial_x^2 X}{X} + 2 \cdot \left(\frac{\partial_x X}{X} \right)^2 = a^2 = \frac{\partial_y^2 Y}{Y} - 2 \cdot \left(\frac{\partial_y Y}{Y} \right)^2$$

$$\Rightarrow 0 = a^2 \cdot X^2 + X \cdot \partial_x^2 X - 2 \cdot (\partial_x X)^2; \quad a^2 \cdot Y^2 + 2 \cdot (\partial_y Y)^2 - Y \cdot \partial_y^2 Y = 0. \tag{434}$$

Solving the "governing" equations gives us the solutions

$$X[x] = \begin{pmatrix} C_x \cdot sech\,[a \cdot (x - x_i)] \\ C_x \cdot sec\,[a \cdot (x - x_i)] \end{pmatrix}; \quad Y[y] = \begin{pmatrix} C_y \cdot sec\,[a \cdot (y - y_i)] \\ C_y \cdot sec\,h\,[a \cdot (y - y_i)] \end{pmatrix}.$$
(435)

Here the C_k are constants, the x_i, y_i, z_i are constant offsets or center points for the sekans hyperbolicus sech respectively the sekans functions sec.

As we have seen before (e.g. [27]) that a zero can still be something, we might also consider the case of an intelligent zero and reinterpret the a-setting above as follows:

$$0 = a_n^2 - a_n^2 = \left(\frac{\partial_x^2 X}{X} + \frac{\partial_y^2 Y}{Y} \right) - 2 \cdot \left(\left(\frac{\partial_x X}{X} \right)^2 + \left(\frac{\partial_y Y}{Y} \right)^2 \right)$$

$$\Rightarrow a_n^2 - \frac{\partial_x^2 X}{X} + 2 \cdot \left(\frac{\partial_x X}{X} \right)^2 = 0 = a_n^2 + \frac{\partial_y^2 Y}{Y} - 2 \cdot \left(\frac{\partial_y Y}{Y} \right)^2$$

$$\Rightarrow 0 = a_n^2 \cdot X^2 + X \cdot \partial_x^2 X - 2 \cdot (\partial_x X)^2; \quad a_n^2 \cdot Y^2 + 2 \cdot (\partial_y Y)^2 - Y \cdot \partial_y^2 Y = 0$$
(436)

which makes it clear that the typical separation approach is just an intelligent-zero technique.

The 3D Case

Now things are getting more interesting because the Einstein tensor does not give automatically zero as it is the case in two dimensions. Again we start with a metric of the form

$$g_{\sigma\kappa} = \begin{pmatrix} F\,[x, y, z] & 0 & 0 \\ 0 & F\,[x, y, z] & 0 \\ 0 & 0 & F\,[x, y, z] \end{pmatrix}.$$
(437)

The separation approach $F\,[x, y, z] = X[x] \cdot Y[y] \cdot Z[z]$ set into (402) gives (415) and again applying true eigenvalues $\tilde{G}_n \equiv F \cdot \hat{G}_n$ we obtain

$$\tilde{G}_n \equiv \tilde{G}_{xn} + \tilde{G}_{yn} + \tilde{G}_{zn} = \frac{3X'[x]^2}{X[x]^2} + \frac{3Y'[y]^2}{Y[y]^2} + \frac{3Z'[z]^2}{Z[z]^2}$$
$$- \frac{2X''[x]}{X[x]} - \frac{2Y''[y]}{Y[y]} - \frac{2Z''[z]}{Z[z]}.$$
(438)

The solutions are all of the kind

$$X[x] = \left(C_x \cdot \sec h^2 \left[\sqrt{\tilde{G}_{xn}} \cdot (x - x_i) \right] \right)$$

$$Y[y] = \left(C_y \cdot \sec h^2 \left[\sqrt{\tilde{G}_{yn}} \cdot (y - y_i) \right] \right)$$

$$Z[z] = \left(C_z \cdot \sec h^2 \left[\sqrt{\tilde{G}_{zn}} \cdot (z - z_i) \right] \right) \tag{439}$$

and we recognize their clear particle character, at least for positive eigenvalues $\tilde{G}_{x,y,z|n}$.

The 4D Case

This time our approach is $F[t, x, y, z] = T[t] * X[x] * Y[y] * Z[z]$, which has to be set into equation (406) and the resulting solution, if following the procedure shown in the 3D cases, would be

$$T[t] = \left(C_t \cdot \exp \left[-\frac{\tilde{G}_{tn}}{2} \cdot t^2 + \mu_t \cdot t \right] \right);$$

$$X[x] = \left(C_x \cdot \exp \left[-\frac{\tilde{G}_{xn}}{2} \cdot x^2 + \mu_x \cdot x \right] \right)$$

$$Y[y] = \left(C_y \cdot \exp \left[-\frac{\tilde{G}_{yn}}{2} \cdot y^2 + \mu_y \cdot y \right] \right);$$

$$Z[z] = \left(C_z \cdot \exp \left[-\frac{\tilde{G}_{zn}}{2} \cdot z^2 + \mu_z \cdot z \right] \right), \tag{440}$$

which are particle solutions, too (for positive eigenvalues $\tilde{G}_{t,x,y,z|n}$), with all sorts of options for spin, iso-spin, matter and antimatter, offset, and subsequent asymmetries.

Summing the Last Section Up

We have a few interesting things here:

(A) From the investigated number of dimensions, only in 4 dimensions we obtain solutions where negative eigenvalues $\tilde{G}_{t,x,y,z|n}$ do not lead to a multitude of singularities but only to an infinite growth with the coordinates going to plus or minus infinity. One can easily check this by setting negative $\tilde{G}_{t,x,y,z|n}$ into the solutions (439) and (440). While there is only a change of sign in (440) (4D case), we go

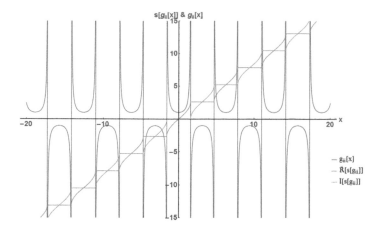

Figure 51 Example for a sekans dependency on one of the spatial coordinates (here x). With the metric components also determining the total curvature, it is clear that a space with such components must curl up, divide and quasi substructure itself. Again, we have also evaluated the real and the imaginary part of the integrated line-element, which is still giving us (no matter the singularities) a stepwise "forward moving" of proper space and proper time (at least as long as we want to see the imaginary portion of the line element as such).

from sekans hyperbolicus to sekans in the 3D case, leading to a multitude of singularities with the coordinates. Only for illustration, we have evaluated such a dependency with respect to the coordinate x assuming a negative eigenvalue, which is leading to $\sec h \left[x \cdot \sqrt{-\tilde{G}_{x|n}} \right]$

$$= \sec \left[x \cdot \sqrt{|\tilde{G}_{x|n}|} \right] \text{ (see Fig. 51)}.$$

(B) All solutions contain probability distributions. Either in the sech form in 2D and 3D (c.f. (435) and (439)) or as Gaussian distributions curves in the 4D case (see (440)). In the latter case, a simple transformation can bring the distributions into normal form. So, taking the $X[x]$ function as example we see that we can reorder as follows:

$$X[x] = \left(C_x \cdot \exp\left[-\frac{\tilde{G}_{xn}}{2} \cdot x^2 + \mu_x \cdot x \right] \right)$$

$$= C_x \cdot e^{-\frac{\tilde{G}_{xn}}{2} \cdot x^2 + \mu_x \cdot x + \frac{2}{\tilde{G}_{xn}} \cdot \left[\left(\frac{\mu_x}{2} \right)^2 - \left(\frac{\mu_x}{2} \right)^2 \right]} = C_x \cdot e^{\frac{\mu_x^2}{2 \cdot \tilde{G}_{xn}}} \cdot e^{-\left(\sqrt{\frac{\tilde{G}_{xn}}{2}} \cdot x - \frac{\mu_x}{\sqrt{2 \cdot \tilde{G}_{xn}}} \right)^2}$$

$$= C_x \cdot e^{\frac{\mu_x^2}{2 \cdot \tilde{G}_{xn}}} \cdot e^{-\frac{\tilde{G}_{xn}}{2} \cdot \left(x - \frac{\mu_x}{\tilde{G}_{xn}} \right)^2},$$

which directly gives us the expectation value x via the integral $C_x \cdot e^{\frac{\mu_x^2}{2 \cdot \tilde{G}_{xn}}}$.

$$\int\limits_{-\infty}^{+\infty} x \cdot e^{-\frac{\tilde{G}_{xn}}{2} \cdot \left(x - \frac{\mu_x}{\tilde{G}_{xn}}\right)^2} dx = \frac{\mu_x}{\tilde{G}_{xn}}; \text{ with } C_x \cdot e^{\frac{\mu_x^2}{2 \cdot \tilde{G}_{xn}}} = \sqrt{\frac{\tilde{G}_{xn}}{2 \cdot \pi}}, \text{ the variance of}$$

the probability distribution $C_x \cdot e^{\frac{\mu_x^2}{2 \cdot \tilde{G}_{xn}}} \cdot \int\limits_{-\infty}^{+\infty} x^2 \cdot e^{-\frac{\tilde{G}_{xn}}{2} \cdot \left(x - \frac{\mu_x}{\tilde{G}_{xn}}\right)^2} dx = \frac{\tilde{G}_{xn} + \mu_x^2}{\tilde{G}_{xn}^2}$

and its standard deviation $\sigma = \frac{1}{\tilde{G}_{xn}}$. We need to note that with the setting

of $C_x \cdot e^{\frac{\mu_x^2}{2 \cdot \tilde{G}_{xn}}} = \sqrt{\frac{\tilde{G}_{xn}}{2 \cdot \pi}}$ we have normalized our solution $X[x]$ such that an

integration of the form $C_x \cdot e^{\frac{\mu_x^2}{2 \cdot \tilde{G}_{xn}}} \cdot \int\limits_{-\infty}^{+\infty} e^{-\frac{\tilde{G}_{xn}}{2} \cdot \left(x - \frac{\mu_x}{\tilde{G}_{xn}}\right)^2} dx$ gives one. This way

we keep the comparability with the classical quantum theory.

(C) From the results obtained so far, one might deduce that one could keep the probability interpretation and handling of the quantum theory, which—altogether—is not too surprising, because we are considering physics on a wiggly space-time. So, we have to note that the possibility of complex μ brings our solutions above directly into the form

$$X[x] = C_x \cdot e^{-\frac{\tilde{G}_{xn}}{2} \cdot x^2 + (\Re[\mu_x] + i \cdot \Im[\mu_x]) \cdot x} = C_x \cdot e^{i \cdot \Im[\mu_x] \cdot x} \cdot e^{\frac{\Re[\mu_x]^2}{2 \cdot \tilde{G}_{xn}}} \cdot e^{-\frac{\tilde{G}_{xn}}{2} \cdot \left(x - \frac{\Re[\mu_x]}{\tilde{G}_{xn}}\right)^2},$$

where we easily recognize the wiggling, respectively, uncertainty-creating, oscillation term $e^{i \cdot \Im[\mu_x] \cdot x}$.

(D) Last but not least, directly following from our considerations in the points above we need an amendment to Heisenberg's uncertainty principle [37]. As this uncertainty principle can be derived from the Cauchy–Schwarz or the Hölder inequality, there must also be a formal way to derive its metric amendment. But as such an investigation should be given some space, we will perform it later in due course of our little "Einstein had it" survey [38]. Here we only give a brief result directly following from solution (440). Let us pick the coordinates x and y, where we are interested in their very principle inequality. Evaluation of the variance and application of the Hölder inequality gives us

$$\Delta x \cdot \Delta y \geq \iiint\limits_{V} \int F \cdot x \cdot y \cdot \mathbf{dx}^4$$

$$= C_x C_y \int\limits_{-\infty}^{+\infty} \int\limits_{-\infty}^{+\infty} x \cdot e^{-\frac{\tilde{G}_{xn}}{2} \cdot \left(x - \frac{\mu_x}{\tilde{G}_{xn}}\right)^2} y \cdot e^{-\frac{\tilde{G}_{yn}}{2} \cdot \left(y - \frac{\mu_y}{\tilde{G}_{yn}}\right)^2} dxdy = \frac{\mu_x}{\tilde{G}_{xn}} \frac{\mu_y}{\tilde{G}_{yn}}.$$

We see that the term $\frac{\mu_x}{\tilde{G}_{xn}} \frac{\mu_y}{\tilde{G}_{yn}}$ must have dimension of length squared. In the Heisenberg picture the length limit would be the Planck length with

$l_P = \sqrt{\frac{\hbar \cdot G}{c^3}} \approx 1.616229 \cdot 10^{-35}$ m. Thus, we have $\left[\frac{\mu_x}{\tilde{G}_{xn}} \frac{\mu_y}{\tilde{G}_{yn}}\right]_{\text{Heisenberg}} =$ $\left[\frac{\mu_x}{\tilde{G}_{xn}} \frac{\mu_y}{\tilde{G}_{yn}}\right]_H = \frac{\hbar \cdot G}{c^3}$. We explicitly point out that this does only hold for the classical Heisenberg principle not for its general metric amendment, but it will suffice in order to show the connection with the classical principle. We have tried to make this clear by the index "H." Now we want to derive the classical Heisenberg uncertainty principle and remember that the momentum p is just the time derivative of a spatial coordinate times mass m. Thus, we can write

$$\Delta x \cdot \Delta p = \Delta x \cdot r_s \cdot \Delta \dot{x} = \Delta x \cdot r_s \cdot \Delta \left(\frac{\partial x}{c \cdot \partial t}\right)$$

and keep the geometrical form and interpretation by using the Schwarzschild radius equivalent for the mass m. Applying our results from above, we have $\Delta x \cdot \Delta p = \Delta x \cdot r_s \cdot \Delta \left(\frac{\partial x}{c \cdot \partial t}\right) \geq \left[\frac{\mu_x}{\tilde{G}_{xn}} \frac{\mu_y}{\tilde{G}_{yn}}\right]_H = \frac{\hbar \cdot G}{c^3}$. Setting the equation for the Schwarzschild radius $r_s = \frac{2 \cdot m \cdot G}{c^2}$ into our result, finally gives us the Heisenberg principle $\Delta x \cdot r_s \cdot \Delta \left(\frac{\partial x}{c \cdot \partial t}\right) \geq \frac{\hbar \cdot G}{c^3} \Rightarrow$ $\Delta x \cdot \Delta \left(\frac{\partial x}{\partial t}\right) \geq \frac{\hbar}{2 \cdot m}$, where we clearly see that the mass matters with respect to resolution achievable regarding certain observables. Taking the more general form from above $\Delta x \cdot \Delta y \geq \frac{\mu_x}{\tilde{G}_{xn}} \frac{\mu_y}{\tilde{G}_{yn}}$, we have to conclude that the curvature of space time influences the principle quantum uncertainty. We see that bigger curvatures (more energy and/or mass) allow for better resolution due to lower uncertainties. This is of interest, especially in connection with the discussion about massive objects like black holes and their "no hair theorem." In one of the next parts of our "Einstein had it" survey this will therefore be investigated in some more detail.

Example: Symmetry of Revolution

For entertainment and illustration, we are now applying a separation approach with the following metric:

$$g_{\sigma\kappa} = \begin{pmatrix} -c^2 \cdot F\,[t, r, \vartheta, \varphi] & 0 & 0 & 0 \\ 0 & F\,[t, r, \vartheta, \varphi] & 0 & 0 \\ 0 & 0 & r^2 \cdot F\,[t, r, \vartheta, \varphi] & 0 \\ 0 & 0 & 0 & r^2 \cdot \sin[\vartheta]^2 \cdot F\,[t, r, \vartheta, \varphi] \end{pmatrix}.$$

$$(441)$$

We apply $F\,[t, r, \vartheta, \varphi] = F\,[r] \cdot \frac{Y[\vartheta,\varphi]}{\sin^2[\vartheta]}$, which leads to

$$
g_{\sigma\kappa} = \begin{pmatrix} -c^2 \cdot F\,[r] \cdot \frac{Y[\vartheta,\varphi]}{\sin^2[\vartheta]} & 0 & 0 & 0 \\ 0 & F\,[r] \cdot \frac{Y[\vartheta,\varphi]}{\sin^2[\vartheta]} & 0 & 0 \\ 0 & 0 & r^2 \cdot F\,[r] \cdot \frac{Y[\vartheta,\varphi]}{\sin^2[\vartheta]} & 0 \\ 0 & 0 & 0 & r^2 \cdot F\,[r] \cdot Y\,[\vartheta, \varphi] \end{pmatrix}
$$

$$(442)$$

and with (402) it gives us the equations

$$
\widehat{G}_n = \widehat{g}_{\,n|\kappa\lambda}\widehat{g}_n^{\kappa\lambda} = \frac{\widehat{G}_{nr}\sin^2[\vartheta] + \widehat{G}_{n\vartheta}\sin[\vartheta]}{F\,[r] \cdot Y\,[\vartheta, \varphi]}
$$

$$
= \widehat{g}_{\,n|\kappa\lambda}\left[g^{\alpha\beta}\begin{pmatrix} -\left[g^{\sigma\kappa}\Gamma^\lambda_{\beta\sigma}\right]_{,\alpha} + \left[g^{\sigma\kappa}\Gamma^\lambda_{\alpha\beta}\right]_{,\sigma} \\ -\Gamma^\mu_{\sigma\alpha}g^{\sigma\kappa}\Gamma^\lambda_{\beta\mu} - \Gamma^\sigma_{\beta\mu}g^{\mu\kappa}\Gamma^\lambda_{\sigma\alpha} \\ +\Gamma^\sigma_{\alpha\beta}g^{\rho\kappa}\Gamma^\lambda_{\sigma\rho} + \Gamma^\rho_{\sigma\rho}g^{\sigma\kappa}\Gamma^\lambda_{\alpha\beta} \end{pmatrix}\right]
$$

$$
\Rightarrow \widehat{G}_{nr} = -\frac{\left(-3r^2 F'[r]^2 + rF\,[r]\,(4F'\,[r] + 3rF''\,[r])\right)}{r^2 F\,[r]^2}
$$

$$
F\,[r] = C_r \cdot e^{-\frac{\widehat{G}_{nr}r^2}{14} - \frac{3\cdot\mu_r}{r^{1/3}}}. \tag{443}
$$

We have only extracted and solved the radial part, because so far we found no way to also solve the angular equations. This is slightly annoying, of course, because we expect to find the elementary particles hidden in those equations. Especially we suspect to find a solution with an angular dependency resembling the hadrons, where centers of excitation (the quarks), in certain angular distributions are forming one hadron. Then, the funny confinement of the quarks would simply be a result of their angular distribution around a certain center point. Trials to separate them only results in adding more excitation energy and thus, forming more particles.

Without the angular solution, we start to concentrate on the radial part alone. Again, we obtain a solution with the Gaussian distribution curve (last equation in (443)), which makes it clear that any radial position can only be given with a certain probability. We should emphasize that here the probability result comes directly from the extended-Einstein field equations. There was no need to do any explicit quantization of the latter. We will discuss further results coming out of this solution in Part III of our "Einstein had it" journey [39].

References

1. T. Bodan, *The Eighth Day*, www.amazon.com/dp/B015R1JPZ2.

2. S. Stamler, Sch. Stamler, T. Bodan, About a multi-scale and multi-fractal universe, in *The Eighth Day* by T. Bodan, ASIN B019M9ZHIE, www.amazon.com/dp/B019M9ZHIE.

3. A. Einstein, Die Grundlage der allgemeinen Relativitätstheorie, *Ann. Phys.* (1916), 354, 769–822.

4. D. Hilbert, Die Grundlagen der Physik, Teil 1, *Göttinger Nachrichten*, (1915), 395–407.

5. P. A. M. Dirac, *The Quantum Theory of the Electron*, 1 February 1928. doi:10.1098/rspa.1928.0023.

6. N. Schwarzer, *Quantized GTR: Understanding the Friedmanns?*, www.amazon.com/dp/B01BQUTKG2.

7. N. Schwarzer, *Understanding the Electron?*, Part I, e-book on www.amazon.com, ASIN B01CCRU6Q6, www.amazon.com/dp/B01CCRU6Q6.

8. T. Bodan, N. Schwarzer, *Quantum Economy*, www.amazon.com/dp/B01N80I0NG.

9. N. Schwarzer, *General Quantization of Smooth Spaces: From ℏ to Plancktensor & Planckfunction*, www.amazon.com/dp/B01N9UGFX2.

10. N. Schwarzer, *Understanding the Electron II*, www.amazon.com/dp/B01N7QUTUL.

11. N. Schwarzer, *The Photon*, www.amazon.com/dp/B06XGC4NDM.

12. N. Schwarzer, *General Quantum Relativity*, www.amazon.com/dp/B01FG5RC0E.

13. N. Schwarzer, *Quantum Tribology, Part I: Theory*, www.amazon.com/dp/B01CI4BI2E.

14. N. Schwarzer, *Understanding Time*, www.amazon.com/dp/B01FNZUVT6.

15. N. Schwarzer, *Recipe to Quantize the General Theory of Relativity*, www.amazon.com/dp/B01FNZUVT6.

16. N. Schwarzer, *The Einstein Oscillator in 1D*, www.amazon.com/dp/B01N5UG8RJ.

17. N. Schwarzer, *Quantized Schwarzschild*, www.amazon.com/dp/B01N7YT6OF.

18. R. P. Kerr, Gravitational field of a spinning mass as an example of algebraically special metrics, *Phys. Rev. Lett.* (1963), 11, 26.

19. R. P. Kerr, Gravitational collapse and rotation, quasi-stellar sources and gravitational collapse, including the *Proceedings of the First Texas Symposium on Relativistic Astrophysics*, edited by I. Robinson, A. Schild, E. L. Schucking. The University of Chicago Press, Chicago and London, (1965), 99–109.

20. Exclusion principle and quantum mechanics, Nobel Lecture, 13 December, 1946. Nobelprize.org [online], (1946).

21. C. Cohen-Tannoudji, B. Diu, F. Laloë, *Quantenmechanik 1&2*, 2. Auflage, Walter de Gruyter, Berlin - New York, (1999).

22. N. Schwarzer, *Quantized Relativized Theology: Where is God?*, www.amazon.com/dp/B01M0XPXTT.

23. N. Schwarzer, *Unorthodox Usage of Higgs Field Mechanism in Every Day Problems*, www.siomec.de/higgs.

24. C. L. Bennett et al., *Nine-Year Wilkinson Microwave Anisotropy Probe (WMAP) Observations: Final Maps and Results*, (2013) arxiv.org/pdf/1212.5225.pdf.

25. G. Nordström, On the energy of the gravitation field in Einstein's theory, *Koninklijke Nederlandse Akademie van Weteschappen Proceedings Series B Physical Sciences*, (1918), 20, 1238–1245.

26. H. Reissner, Über die Eigengravitation des elektrischen Feldes nach der Einsteinschen Theorie, *Ann. Phys.* (1916), 355(9), 106–120.

27. N. Schwarzer, *Our Universe, Nothing but an Intelligent Zero?: The Dark Lord's Zero-Sum- & God's Non-Zero-Sum-Game*, www.amazon.com/dp/B072J9H1BY.

28. G. W. Gibbons, S. W. Hawking, Cosmological event horizons, thermodynamics, and particle creation, *Phys. Rev. D*, (1977), 15, 2738.

29. Z. Stuchlik, M. Calvani, Null geodesics in black hole metrics with non-zero cosmological constant, *Gen. Rel. Grav.* (1991), 23, 507.

30. A. Müller, www.spektrum.de/astrowissen/lexdt_k02.html.

31. N. Schwarzer, *How Einstein Gives Dirac, Klein-Gordon and Schrödinger*, www.amazon.com/dp/B071K2Y4V2.

32. N. Schwarzer, *Einstein Already had it, But He Did not See it, Part IV: Sixty e-Foldings*, www.amazon.de.

33. E. Witten, Anti-de sitter space and holography, *Adv. Theor. Math. Phys.* (1998), 2, 253–291, arxiv:hep-th/9802150.

34. G. Hooft, *Dimensional Reduction in Quantum Gravity*, (1993), arxiv:gr-qc/9310026; *The Holographic Principle*, arxiv:hep-th/0003004.

35. L. Susskind, The world as a hologram, *J. Math. Phys.*, (1995), 36, 6377, arxiv:hep-th/9409089.

36. K. Schwarzschild, On the gravitational field of a mass point according to Einstein's theory (translation and foreword by S. Antoci and A. Loinger), *Sitzungsber. Preuss. Akad. Wiss. Berlin (Math. Phys.)*, (1916), 1916, 189–196, arXiv:physics/9905030v1.

37. W. Heisenberg, Über den anschaulichen Inhalt der quantentheoretischen Kinematik und Mechanik, *Z. Phys.* (1927), 43(3–4), 172–198.

38. N. Schwarzer, *Einstein had it, But He Did not See it, Part V: Amendment to the Heisenberg Uncertainty Principle*, www.amazon.com/dp/B074MB3J3S.

39. N. Schwarzer, *Einstein had it, But He Did not See it, Part III: The Impossible Black-Hole Singularity*, www.amazon.com/dp/B074LV1RPD.

40. N. Schwarzer, *Einstein had it, But He Did not See it, Part XX: Higher Order Covariant Variation of the Einstein-Hilbert-Action*, www.amazon.com/dp/B0788VWKD4.

Index